# The Essential Welding Pocket Reference

## A Field Guide to Metals, Filler Materials, and Modern Welding Processes

By **Jim Cox**

**DISCLAIMER**

This book is intended as a **quick-reference guide** for welders and metalworkers. Welding involves heat, electricity, pressurized gases, hazardous fumes, and other risks that can cause injury, death, or property damage.

The information in this guide is provided **for educational purposes only**.
Real-world welding conditions vary, and the user is responsible for:

- Following all OSHA, AWS, and local safety regulations
- Using proper PPE and ventilation
- Verifying settings and procedures with equipment manufacturers
- Exercising professional judgment in all welding operations

The author and publisher assume **no liability** for accidents, injuries, or damages resulting from the use or misuse of the information contained in this book.

# Table of Contents

**Section 7: Reference Charts & Appendices**

## INTRODUCTION

Welding is one of the few trades where **your hands, your eyes, and your judgment** matter more than the machine in front of you. Whether you're burning rods on a farm gate, TIG-welding aluminum, or running flux-core on structural steel, the job demands **speed, accuracy, and confidence**.

This guide is built for welders who don't have time to dig through manuals, search online, or guess at settings. It's designed to live in your pocket, your toolbox, or your truck—ready the moment you need it.

Inside, you'll find:

- **Metals broken down simply**—how to identify them, what they're used for, and how they behave under heat
- **Filler metal selection**—rods, wires, and TIG fillers matched to the right metals
- **Settings charts**—amps, volts, wire speed, polarity, and gas flow
- **Troubleshooting pages**—fast fixes for real-world problems
- **Symbols and joint prep**—the essentials without the fluff

This is the book welders reach for when they need answers **right now**.

## HOW TO USE THIS GUIDE

This pocket reference is built for **speed**. Every section is structured so you can flip to what you need and get back to work.

### 1. Identify the Metal

Go to **Part I — Metals Encyclopedia**.
Each metal includes:

- How to identify it
- Weldability rating
- Preheat/interpass guidance
- Best processes
- Best filler metals

### 2. Choose the Process

Stick, MIG, flux-core, or TIG—each has its own strengths.
Use **Part II** to match the filler to the metal.

### 3. Set the Machine

Go to **Part III** for:

- Amps
- Volts
- Wire feed speed
- Gas flow
- Polarity
- Tungsten size

Settings are given in **practical ranges**, not lab-perfect numbers.

### 4. Fix Problems Fast

Part IV gives you:

- What the defect looks like
- Why it happens
- How to fix it immediately

### 5. Use the Quick Reference Pages

These are the "no-thinking-required" charts welders love:

- What filler to use on what metal
- What process works best
- Settings at a glance

## SECTION 1 — METALS & METALLURGY BASICS

### CHAPTER 1 — UNDERSTANDING STEEL TYPES

*(Carbon steel, alloy steel, structural grades, behavior under heat)*

Steel is the backbone of welding. Understanding how it behaves under heat, stress, and filler metal selection is the foundation of every weld.

#### 1. Carbon Steel Categories

Steel is classified by carbon content:

| Type | Carbon % | Characteristics | Common Uses |
|---|---|---|---|
| **Low Carbon (Mild Steel)** | 0.05–0.30% | Soft, ductile, easy to weld | Structural, tubing, plate |
| **Medium Carbon** | 0.30–0.60% | Stronger, harder | Shafts, axles |
| **High Carbon** | 0.60–1.0% | Hard, brittle | Springs, knives |
| **Ultra-High Carbon** | 1.0%+ | Very hard | Tooling |

Low-carbon steel is the welder's bread and butter.

**2. Structural Steel Grades**

Common structural steels:

- **A36** — most common plate steel
- **A500** — tubing
- **A572** — high-strength structural
- **A514 (T-1)** — quenched & tempered, heat-sensitive

**3. Alloy Steels**

Alloy steels contain chromium, molybdenum, nickel, vanadium, etc.

Examples:

- **4130/4140** — chromoly
- **8620** — case-hardening steel
- **6150** — spring steel

These steels often require:

- Preheat
- Controlled cooling
- Correct filler (ER80S-D2, ER312, etc.)

**4. Weldability Factors**

Steel weldability depends on:

- Carbon content
- Alloy content
- Thickness
- Cooling rate
- Preheat/interpass temperature

The higher the carbon, the more crack-sensitive the steel.

## CHAPTER 2 — STAINLESS STEEL FAMILIES

*(Austenitic, ferritic, martensitic, duplex, precipitation-hardening)*

Stainless steel is defined by **10.5%+ chromium**, which forms a corrosion-resistant oxide layer.

### 1. Austenitic Stainless (300 Series)

- **Non-magnetic**
- **Most weldable**
- **Most common**

Examples:

- **304** — general purpose
- **316** — marine grade
- **309** — dissimilar metals

Filler metals:

- ER308L, ER316L, ER309L

### 2. Ferritic Stainless (400 Series)

- Magnetic
- Lower corrosion resistance
- Used in automotive exhaust

Examples:

- **409, 430**

**3. Martensitic Stainless**

- Hard, strong
- Heat-treatable
- Crack-sensitive

Examples:

- **410, 420**

**4. Duplex Stainless**

- Very strong
- Used in chemical plants
- Requires controlled heat input

**5. Welding Considerations**

- Avoid overheating
- Use proper filler
- Back-purge pipe
- Control interpass temperature

## CHAPTER 3 — ALUMINUM ALLOYS

*(1xxx–7xxx series, weldability, filler selection)*

Aluminum is lightweight, corrosion-resistant, and highly conductive.

### 1. Aluminum Alloy Series

| Series | Characteristics | Weldability |
|---|---|---|
| **1xxx** | Pure aluminum | Excellent |
| **3xxx** | Manganese alloys | Excellent |
| **5xxx** | Magnesium alloys | Excellent |
| **6xxx** | Magnesium + silicon | Good |
| **7xxx** | Zinc alloys | Poor |

### 2. Common Alloys

- **3003** — sheet metal
- **5052** — marine aluminum
- **6061** — structural aluminum
- **7075** — aircraft (not weldable)

### 3. Filler Selection

- **4043** — easy flow
- **5356** — stronger, stiffer

**4. Welding Considerations**

- Remove oxide layer
- Use AC TIG or MIG with argon
- Preheat thick sections
- Avoid contamination

## CHAPTER 4 — CAST IRON TYPES & BEHAVIOR

*(Gray, ductile, malleable, white iron)*

Cast iron is brittle, carbon-rich, and crack-sensitive.

### 1. Types of Cast Iron

| Type | Characteristics | Weldability |
|---|---|---|
| **Gray Iron** | Flake graphite | Fair |
| **Ductile Iron** | Spherical graphite | Good |
| **Malleable Iron** | Heat-treated | Fair |
| **White Iron** | Hard, brittle | Not weldable |

### 2. Filler Metals

- **Ni-99** — machinable
- **Ni-55** — stronger
- **Silicon bronze** — low heat

### 3. Welding Rules

- Preheat 300–600°F
- Short beads
- Peen lightly
- Slow cool

## CHAPTER 5 — TOOL STEELS & HARDENING

*(W-series, O-series, A-series, D-series, H-series)*

Tool steels are designed for hardness, wear resistance, and strength.

### 1. Tool Steel Families

| Series | Meaning | Notes |
|---|---|---|
| **W-Series** | Water-hardening | Do not weld |
| **O-Series** | Oil-hardening | Preheat required |
| **A-Series** | Air-hardening | Very crack-sensitive |
| **D-Series** | High carbon | Often brazed |
| **H-Series** | Hot-work steels | High heat resistance |

### 2. Welding Considerations

- Preheat 400–1000°F
- Use ER312 or Ni-based fillers
- Slow cool
- Expect hardness changes

## CHAPTER 6 — HIGH-STRENGTH STEELS (HSLA)

*(A572, A514, AR plate, quenched & tempered steels)*

HSLA steels offer high strength with lower weight.

### 1. Common Grades

- **A572** — structural
- **A514 (T-1)** — quenched & tempered
- **AR Plate** — abrasion-resistant

### 2. Welding Rules

- Preheat 150–400°F
- Control interpass temperature
- Avoid overheating T-1
- Use proper filler (E80, E90, ER80S-D2)

## CHAPTER 7 — NON-FERROUS METALS

*(Copper, brass, bronze)*

These metals conduct heat extremely well and require special technique.

### 1. Copper

- Very conductive
- Requires high heat
- Use silicon bronze or ERCu filler

### 2. Brass

- Copper + zinc
- Zinc fumes are hazardous
- TIG with silicon bronze

### 3. Bronze

- Copper + tin
- Stronger than brass
- TIG with ERCuSi-A

## CHAPTER 8 — TITANIUM & EXOTIC ALLOYS

*(Titanium, Inconel, Monel, Hastelloy)*

These metals require extreme cleanliness and shielding.

### 1. Titanium

- Lightweight, strong
- Must be welded in **argon bubble**
- Use CP-Ti or Ti-6Al-4V filler

### 2. Nickel Alloys

Examples:

- **Inconel**
- **Monel**
- **Hastelloy**

Characteristics:

- High heat resistance
- Corrosion-proof
- Used in aerospace and chemical plants

**3. Welding Rules**

- Perfect cleanliness
- Gas lens
- Long post-flow
- No drafts

## CHAPTER 9 — HEAT TREATMENT & METALLURGY BASICS

*(Annealing, normalizing, quenching, tempering)*

Heat treatment changes metal hardness, strength, and ductility.

### 1. Annealing

- Softens metal
- Improves machinability

### 2. Normalizing

- Refines grain structure
- Improves toughness

### 3. Quenching

- Rapid cooling
- Increases hardness
- Increases brittleness

### 4. Tempering

- Reheating quenched steel
- Reduces brittleness
- Restores toughness

### 5. Welding Effects

Welding creates a **heat-affected zone (HAZ)** that:

- Changes hardness
- Alters grain structure
- Can cause cracking

Preheat and interpass control reduce HAZ problems.

## SECTION 2 — METALS ENCYCLOPEDIA (In-Depth Field Guides)

## CHAPTER 10 — MILD STEEL (A36 & COMMON STRUCTURAL STEELS)

### IDENTIFICATION

**Common names:**
A36, mild steel, low-carbon steel, structural steel, hot-rolled steel.

**How to recognize it:**

- Dull gray surface, often with mill scale
- Magnetic
- Soft compared to alloy steels
- Sparks are bright yellow with branching bursts
- Bends easily, machines easily, welds easily

**Typical uses:**

- Structural beams, angle iron, flat bar
- Farm repairs, gates, trailers
- General fabrication
- Pipe (non-pressure)
- Brackets, frames, supports

**Weldability:** ★★★★★ (Excellent)
Mild steel is the **easiest metal to weld** with any process.

## PREHEAT & INTERPASS GUIDANCE

- **Normal thickness (up to 1"):** No preheat required
- **Cold weather (< 40°F):** Preheat to **100–150°F**
- **Thick sections (> 1"):** Preheat **150–250°F**
- **Interpass max:** ~400°F

**Notes:**

- Preheat helps avoid hydrogen cracking in thick or restrained joints.
- Mill scale should be removed for best results, especially with MIG and TIG.

## BEST PROCESSES FOR MILD STEEL

| Process | Strengths | Weaknesses | Best Use Cases |
|---|---|---|---|
| **Stick (SMAW)** | Works on dirty/rusty steel, outdoor friendly | More spatter, slower | Farm repairs, structural, field work |
| **MIG (GMAW)** | Fast, clean, easy to learn | Needs clean metal, wind sensitive | Shop fabrication, production |
| **Flux-Core (FCAW)** | High penetration, outdoor capable | Slag, more fumes | Structural, heavy plate, outdoor |
| **TIG (GTAW)** | Cleanest, precise | Slow, requires clean metal | Thin sheet, precision work |

## FILLER METAL SELECTION

### Stick Electrodes (SMAW)

| Rod | Use Case | Polarity | Notes |
|---|---|---|---|
| **6010** | Root passes, pipe, dirty steel | DCEP | Deep penetration, fast freeze |
| **6011** | Dirty/rusty steel, AC machines | AC / DCEP | Good for farm repairs |
| **6013** | Thin sheet, smooth bead | AC / DCEP | Low penetration |
| **7014** | Flat/Horizontal, smooth bead | AC / DCEP | Iron powder, fast fill |
| **7018** | Structural, high strength | DCEP | Low-hydrogen, keep dry |
| **7024** | Flat fillets, heavy deposition | AC / DCEP | "Drag rod," very fast |

**Recommended all-around rod: 7018**
**Best for dirty steel: 6011**

**MIG Solid Wire (GMAW)**

| Wire | Gas | Use Case | Notes |
|---|---|---|---|
| **ER70S-6** | 75/25 or $CO_2$ | General fabrication | Most common mild steel wire |
| **ER70S-3** | 75/25 | Clean steel | Slightly less deoxidizers |
| **Metal-core (E70C-6M)** | 90/10 | High deposition | Production, flat/horizontal |

**Recommended all-around wire: ER70S-6**

**Flux-Core Wire (FCAW)**

| Wire | Shielding | Use Case | Notes |
|---|---|---|---|
| **E71T-11** | Self-shielded | Outdoor, dirty steel | Great for field repairs |
| **E71T-1** | Gas-shielded | Structural, shop work | Smooth bead, high strength |
| **E71T-8** | Self-shielded | Heavy structural | Deep penetration, code work |

**TIG Filler Rods (GTAW)**

| Filler | Use Case | Notes |
|---|---|---|
| **ER70S-2** | Dirty or unknown steel | Triple-deoxidized |
| **ER70S-6** | Clean steel | Most common TIG filler |

## SETTINGS & PARAMETERS

These are **real-world ranges**, not lab numbers.

### STICK (SMAW) SETTINGS — MILD STEEL

#### E7018

| Rod Size | Thickness Range | Amps | Notes |
|---|---|---|---|
| 3/32" | 1/8–3/16 | 70–100 | Good for thinner steel |
| 1/8" | 1/8–1/4 | 115–135 | Most common size |
| 5/32" | 1/4–3/8 | 150–190 | Heavy plate |

#### E6011

| Rod Size | Amps | Notes |
|---|---|---|
| 3/32" | 60–90 | Good for thin/dirty steel |
| 1/8" | 85–120 | Deep penetration |
| 5/32" | 110–160 | Heavy rust/paint |

## MIG (GMAW) SETTINGS — ER70S-6, 75/25 GAS

### 0.030" Wire

| Thickness | Volts | WFS (ipm) | Notes |
|---|---|---|---|
| 16 ga | 16–17 | 200–260 | Stitch to avoid burn-through |
| 1/8" | 17–18.5 | 260–320 | Straight or slight weave |
| 3/16" | 18.5–19.5 | 320–380 | Good penetration |
| 1/4" | 20–21 | 380–450 | Multi-pass recommended |

### 0.035" Wire

| Thickness | Volts | WFS (ipm) |
|---|---|---|
| 1/8" | 17–19 | 200–300 |
| 3/16" | 19–20.5 | 300–360 |
| 1/4" | 20.5–22 | 360–420 |

## FLUX-CORE (FCAW) SETTINGS — E71T-11

### 0.035" Wire

| Thickness | Volts | WFS (ipm) | Notes |
|---|---|---|---|
| 1/8" | 18–19 | 250–300 | Outdoor friendly |
| 3/16" | 19–20 | 300–350 | Deep penetration |
| 1/4" | 20–21 | 350–400 | Multi-pass |

## TIG (GTAW) SETTINGS — MILD STEEL

| Thickness | Tungsten | Amps | Gas Flow | Notes |
|---|---|---|---|---|
| 1/16" | 1/16" | 40–70 | 10–15 CFH | ER70S-2 recommended |
| 1/8" | 3/32" | 90–130 | 12–18 CFH | Most common |
| 3/16" | 1/8" | 130–180 | 15–20 CFH | Preheat helps |

## WELDING TECHNIQUE NOTES

### Stick

- Drag 7018; whip/pause 6010/6011
- Keep arc short
- Maintain rod angle ~10–15°

### MIG

- Push for cleaner bead
- Keep stickout 3/8"
- Avoid wind

### Flux-Core

- Drag angle
- Expect slag
- Great for outdoor work

**TIG**

- Tight arc length
- Clean metal
- Use foot pedal or steady amperage control

**COMMON PROBLEMS & FIXES (MILD STEEL)**

| Problem | Likely Cause | Fast Fix |
|---|---|---|
| Porosity | Wind, dirty metal, low gas | Clean metal, increase gas, block wind |
| Undercut | Too hot, too fast | Reduce heat, slow travel |
| Cold lap | Too cold, long stickout | Increase heat, shorten stickout |
| Spatter | Wrong settings, $CO_2$ gas | Adjust volts/WFS, use 75/25 |
| Cracking | Thick steel, cold weather | Preheat 150–250°F |

# CHAPTER 11 — HIGH-STRENGTH LOW-ALLOY STEELS (HSLA)

*(A572, A588, A992, DOM tubing, automotive HSLA, structural grades)*

## IDENTIFICATION

**Common names:**
HSLA, high-strength low-alloy, structural alloy steel, A572, A588 (weathering steel), A992 (-flange beams), automotive HSLA.

**How to recognize it:**

- Looks similar to mild steel
- Often used in **structural beams, heavy equipment, automotive frames, roll cages**
- Slightly harder surface feel
- Sparks similar to mild steel but slightly more “bursting”
- Stronger and lighter than A36 for the same thickness

**Typical uses:**

- Structural beams and columns
- Heavy equipment frames
- Automotive frames, suspension components
- Roll cages (DOM tubing)
- Bridges (A588 weathering steel)
- Crane booms, excavator arms

**Weldability:** ★★★★☆ (Very good, but requires more care than mild steel)
HSLA steels are weldable, but **heat input control** matters to avoid weakening the heat-affected zone (HAZ).

## PREHEAT & INTERPASS GUIDANCE

HSLA steels vary widely, but these rules of thumb work in the field:

**General HSLA (A572, A992, automotive HSLA)**

- **Up to 1/2" thick:** Preheat **50–150°F**
- **1/2"–1" thick:** Preheat **150–250°F**
- **Over 1":** Preheat **250–350°F**
- **Interpass max:** ~450°F

**Weathering steel (A588 / COR-TEN)**

- Preheat **100–200°F** depending on thickness
- Avoid excessive heat input to preserve corrosion resistance

**DOM tubing (roll cage material)**

- **No preheat needed** for thin wall
- Avoid overheating — can weaken tube integrity
- TIG or MIG preferred

**Notes:**

- HSLA steels lose strength if overheated
- Avoid weaving with 7018 on thin HSLA — use stringer beads
- Keep interpass temperature controlled

## BEST PROCESSES FOR HSLA

| Process | Strengths | Weaknesses | Best Use Cases |
|---|---|---|---|
| **Stick (SMAW)** | Strong welds, field-ready | Must control heat | Structural, heavy equipment |
| **MIG (GMAW)** | Clean, fast, consistent | Needs clean metal | Automotive, fabrication |
| **Flux-Core (FCAW)** | High strength, outdoor capable | Slag, fumes | Structural, bridges, heavy plate |
| **TIG (GTAW)** | Precise, low heat input | Slow | DOM tubing, thin HSLA |

## FILLER METAL SELECTION

### Stick Electrodes (SMAW)

| Rod | Use Case | Polarity | Notes |
|---|---|---|---|
| **7018** | Structural HSLA | DCEP | Standard choice |
| **8018-C3** | Low-temp toughness | DCEP | For cold climates |
| **9018** | High-strength HSLA | DCEP | Heavy equipment |
| **11018** | Very high strength | DCEP | Crane booms, critical welds |

**Recommended all-around rod: 7018**
**For high-strength structural: 8018 or 9018**

### MIG Solid Wire (GMAW)

| Wire | Gas | Use Case | Notes |
|---|---|---|---|
| **ER70S-6** | 75/25 | General HSLA | Works for most jobs |
| **ER80S-D2** | 75/25 or 90/10 | Higher strength | Great for A572/A588 |
| **ER90S-3** | 90/10 | Heavy structural | Stronger weld metal |

**Recommended for structural HSLA: ER80S-D2**

**Flux-Core Wire (FCAW)**

| Wire | Shielding | Use Case | Notes |
| --- | --- | --- | --- |
| **E71T-1** | Gas-shielded | Structural HSLA | Smooth bead, high strength |
| **E81T-1** | Gas-shielded | Higher strength | Bridges, heavy equipment |
| **E91T-1** | Gas-shielded | Very high strength | Critical structural work |
| **E71T-8** | Self-shielded | Outdoor structural | Deep penetration |

**Recommended for field structural: E71T-8**
**Recommended for shop structural: E81T-1**

**TIG Filler Rods (GTAW)**

| Filler | Use Case | Notes |
| --- | --- | --- |
| **ER70S-2** | General HSLA | Good deoxidizers |
| **ER80S-D2** | Higher strength | Best for A572/A588 |
| **ER90S-B3** | High-temp service | Boilers, pressure parts |

## SETTINGS & PARAMETERS

These are practical field ranges.

### STICK (SMAW) SETTINGS — HSLA

**7018**

| Rod Size | Thickness Range | Amps | Notes |
|---|---|---|---|
| 3/32" | 1/8–3/16 | 80–100 | Good for thin HSLA |
| 1/8" | 1/8–1/4 | 115–135 | Most common |
| 5/32" | 1/4–3/8 | 150–190 | Heavy plate |

**8018 / 9018**

| Rod Size | Amps | Notes |
|---|---|---|
| 1/8" | 120–140 | Slightly hotter than 7018 |
| 5/32" | 150–200 | Structural work |

### MIG (GMAW) SETTINGS — ER80S-D2

**0.035" Wire**

| Thickness | Volts | WFS (ipm) | Notes |
|---|---|---|---|
| 1/8" | 18–19 | 240–300 | Good penetration |
| 3/16" | 19–20.5 | 300–360 | Structural |
| 1/4" | 20.5–22 | 360–420 | Multi-pass |

**0.045" Wire**

| Thickness | Volts | WFS (ipm) |
|---|---|---|
| 3/16" | 20–22 | 200–260 |
| 1/4" | 22–24 | 260–320 |
| 3/8" | 24–26 | 320–380 |

**FLUX-CORE (FCAW) SETTINGS — E81T-1**

**0.045" Wire**

| Thickness | Volts | WFS (ipm) | Notes |
|---|---|---|---|
| 1/4" | 22–24 | 200–260 | Strong, smooth bead |
| 3/8" | 24–26 | 260–320 | Structural |
| 1/2" | 26–28 | 320–380 | Multi-pass |

**TIG (GTAW) SETTINGS — HSLA**

| Thickness | Tungsten | Amps | Gas Flow | Notes |
|---|---|---|---|---|
| 1/16" | 1/16" | 50–80 | 10–15 CFH | Avoid overheating |
| 1/8" | 3/32" | 90–130 | 12–18 CFH | ER80S-D2 recommended |
| 3/16" | 1/8" | 130–180 | 15–20 CFH | Keep heat input low |

## WELDING TECHNIQUE NOTES

### Stick

- Use **stringer beads**, not wide weaves
- Maintain controlled heat input
- Keep rods dry (especially 8018/9018)

### MIG

- Push angle for cleaner bead
- Avoid excessive heat on thin HSLA
- Use proper gas (75/25 or 90/10)

### Flux-Core

- Drag angle
- Great for outdoor structural
- Expect slag; chip thoroughly

### TIG

- Tight arc length
- Use fillers matched to strength
- Avoid overheating thin HSLA

## COMMON PROBLEMS & FIXES (HSLA)

| Problem | Likely Cause | Fast Fix |
|---|---|---|
| Loss of strength | Too much heat | Use stringers, reduce amps |
| Cracking | No preheat, cold weather | Preheat 150–250°F |
| Porosity | Dirty metal, wind | Clean, block wind, adjust gas |
| Undercut | Too hot, too fast | Reduce heat, slow travel |
| HAZ softening | Overheating | Control interpass temp |

## CHAPTER 12 — QUENCHED & TEMPERED STEELS (Q&T STEELS)

*(T1, AR400, AR500, 4130/4140, Hardox, abrasion-resistant plate)*

### IDENTIFICATION

**Common names:**
Q&T steel, quenched and tempered steel, AR plate, T1 steel, Hardox, 4130/4140 chromoly.

**How to recognize it:**

- Harder surface feel than mild or HSLA
- Often used in **wear applications** (buckets, blades, dump beds)
- Sparks are more "bursting" than mild steel
- Rings with a higher pitch when struck
- Often marked with grade (AR400, AR500, T1, etc.)

**Typical uses:**

- Excavator buckets, dozer blades
- Dump truck liners
- Mining equipment
- Wear plates
- High-strength structural components
- Roll cages (4130 chromoly)
- Firearm components (AR500 targets)

**Weldability:** ★★☆☆☆ (Fair, but requires strict control)
Q&T steels **lose strength when overheated**. Heat input control is critical.

**PREHEAT & INTERPASS GUIDANCE**

This is the most important part of welding Q&T steels.

**General Q&T Plate (T1, AR400, AR500)**

- **Up to 1/2" thick:** Preheat **200–300°F**
- **1/2"–1" thick:** Preheat **300–400°F**
- **Over 1":** Preheat **400–500°F**
- **Interpass max: 450°F** (do NOT exceed)

**4130 / 4140 Chromoly**

- **Thin wall tubing (< 0.120")**: Preheat **not required**
- **Thicker sections:** Preheat **300–400°F**
- **Interpass max: 450°F**

**Hardox (manufacturer-specific)**

- Hardox 400/450: Preheat **200–300°F**
- Hardox 500: Preheat **300–400°F**
- Follow manufacturer guidance when possible

**Critical notes:**

- **Never quench after welding** — let it cool slowly
- **Avoid weaving** — use stringer beads only
- **Do not over weld** — excessive heat destroys hardness
- **Use low-hydrogen consumables only**

## BEST PROCESSES FOR Q&T STEELS

| Process | Strengths | Weaknesses | Best Use Cases |
|---|---|---|---|
| **Stick (SMAW)** | Best control of heat, low-hydrogen | Slower | Field repairs, heavy plate |
| **MIG (GMAW)** | Clean, controlled | Must watch heat input | Shop repairs, fabrication |
| **Flux-Core (FCAW)** | Strong welds, good penetration | Heat input can be high | Structural Q&T, heavy equipment |
| **TIG (GTAW)** | Precise, low heat | Slow, requires skill | Chromoly tubing |

## FILLER METAL SELECTION

### Stick Electrodes (SMAW)

| Rod | Use Case | Polarity | Notes |
|---|---|---|---|
| **7018** | General Q&T welding | DCEP | Standard choice |
| **8018-C3** | Higher toughness | DCEP | Good for cold climates |
| **9018** | High-strength Q&T | DCEP | Heavy equipment |
| **11018** | Very high strength | DCEP | Crane booms, critical repairs |

**Recommended all-around rod: 7018**
**For AR plate: 8018-C3** or **9018**

**MIG Solid Wire (GMAW)**

| Wire | Gas | Use Case | Notes |
| --- | --- | --- | --- |
| **ER70S-6** | 75/25 | General repairs | Works but not strength-matched |
| **ER80S-D2** | 75/25 or 90/10 | Higher strength | Best for AR400/T1 |
| **ER90S-3** | 90/10 | High-strength structural | Stronger weld metal |

**Recommended for Q&T steels: ER80S-D2**

**Flux-Core Wire (FCAW)**

| Wire | Shielding | Use Case | Notes |
| --- | --- | --- | --- |
| **E71T-1** | Gas-shielded | General Q&T | Smooth bead |
| **E81T-1** | Gas-shielded | Higher strength | Structural |
| **E91T-1** | Gas-shielded | Very high strength | Heavy equipment |
| **E71T-8** | Self-shielded | Outdoor work | Deep penetration |

**TIG Filler Rods (GTAW)**

| Filler | Use Case | Notes |
|---|---|---|
| **ER70S-2** | General Q&T | Good deoxidizers |
| **ER80S-D2** | Strength-matched | Best for T1/AR400 |
| **ER90S-B3** | High-temp service | Boilers, pressure parts |

## SETTINGS & PARAMETERS

These are practical field ranges.

### STICK (SMAW) SETTINGS — Q&T STEELS

**7018 / 8018 / 9018**

| Rod Size | Thickness Range | Amps | Notes |
|---|---|---|---|
| 3/32" | 1/8–3/16 | 80–100 | Good for controlled heat |
| 1/8" | 1/8–1/4 | 115–135 | Most common |
| 5/32" | 1/4–3/8 | 150–190 | Heavy plate |

**Technique:**

- Use **stringer beads only**
- Keep arc short
- Control interpass temperature

**MIG (GMAW) SETTINGS — ER80S-D2**

**0.035" Wire**

| Thickness | Volts | WFS (ipm) | Notes |
|---|---|---|---|
| 1/8" | 17–18.5 | 220–280 | Keep heat low |
| 3/16" | 18.5–20 | 280–340 | Good penetration |
| 1/4" | 20–21.5 | 340–400 | Multi-pass |

**0.045" Wire**

| Thickness | Volts | WFS (ipm) |
|---|---|---|
| 3/16" | 20–22 | 200–260 |
| 1/4" | 22–24 | 260–320 |
| 3/8" | 24–26 | 320–380 |

**FLUX-CORE (FCAW) SETTINGS — E81T-1**

**0.045" Wire**

| Thickness | Volts | WFS (ipm) | Notes |
|---|---|---|---|
| 1/4" | 22–24 | 200–260 | Strong, smooth bead |
| 3/8" | 24–26 | 260–320 | Structural |
| 1/2" | 26–28 | 320–380 | Multi-pass |

**TIG (GTAW) SETTINGS — 4130/4140**

| Thickness | Tungsten | Amps | Gas Flow | Notes |
|---|---|---|---|---|
| 0.065" | 1/16" | 50–70 | 10–15 CFH | No preheat |
| 0.095" | 3/32" | 70–110 | 12–18 CFH | ER80S-D2 recommended |
| 1/8" | 3/32" | 110–150 | 15–20 CFH | Control heat carefully |

**WELDING TECHNIQUE NOTES**

**Stick**

- Use **stringer beads only**
- Keep heat input low
- Maintain preheat and interpass temps

**MIG**

- Push angle
- Avoid weaving
- Use multiple small passes

**Flux-Core**

- Drag angle
- Excellent for structural Q&T
- Watch heat input

**TIG**

- Tight arc
- Use strength-matched filler
- Avoid overheating chromoly

**COMMON PROBLEMS & FIXES (Q&T STEELS)**

| Problem | Likely Cause | Fast Fix |
|---|---|---|
| Cracking | No preheat, too much heat | Preheat 200–400°F, use stringers |
| Loss of hardness | Overheating | Reduce heat, smaller beads |
| HAZ softening | Excessive interpass temp | Keep below 450°F |
| Porosity | Dirty metal, wind | Clean, block wind, adjust gas |
| Brittle welds | Wrong filler | Use ER80S-D2 or 8018/9018 |

## CHAPTER 13 — STAINLESS STEELS

*(304, 316, 309, 409, 410, 430, duplex overview)*

### IDENTIFICATION

**Common names:**
Stainless steel, SS, 304, 316, 309, 409, 410, 430, food-grade stainless, exhaust stainless.

**How to recognize it:**

- Bright silver appearance
- Non-magnetic (304/316) or slightly magnetic (409/430)
- Does not rust easily
- Produces **short, reddish sparks** (much less than carbon steel)
- Often used where corrosion resistance matters

**Typical uses:**

- Food processing equipment
- Kitchen equipment
- Marine hardware
- Exhaust systems (409, 304)
- Chemical plants
- Dairy equipment
- Handrails, architectural work

**Weldability:** ★★★★☆ (Good, but requires cleanliness and proper filler)
Stainless is easy to weld **if you keep it clean and control heat.**

**STAINLESS STEEL TYPES**

**Austenitic (Non-magnetic)**

**304, 304L, 316, 316L, 309**

- Most common
- Excellent corrosion resistance
- Easy to weld
- Low carbon ("L") grades reduce carbide precipitation

**Ferritic (Magnetic)**

**409, 430**

- Cheaper stainless
- Used in automotive exhaust
- Lower corrosion resistance
- Weldable but heat-sensitive

**Martensitic (Magnetic, hardenable)**

**410, 420**

- High hardness
- Used for knives, tools
- Requires preheat and post-heat
- More crack-sensitive

**Duplex (Advanced)**

**2205, 2507**

- High strength
- Corrosion resistant
- Requires strict heat control
- Not common in general fab shops

**PREHEAT & INTERPASS GUIDANCE**

**Austenitic (304/316/309)**

- **No preheat required**
- Keep interpass **below 350°F**
- Avoid overheating to prevent warping and carbide precipitation

**Ferritic (409/430)**

- Preheat **150–300°F**
- Interpass **below 300°F**
- Avoid rapid cooling

**Martensitic (410/420)**

- Preheat **400–600°F**
- Interpass **below 600°F**
- Slow cool after welding
- Use matching filler or 309 for crack resistance

## BEST PROCESSES FOR STAINLESS

| Process | Strengths | Weaknesses | Best Use Cases |
|---|---|---|---|
| **TIG (GTAW)** | Cleanest, precise | Slow | Food-grade, thin sheet |
| **MIG (GMAW)** | Fast, clean | Needs clean metal | Fabrication, exhaust |
| **Stick (SMAW)** | Good for field work | Slag, slower | Repairs, outdoor |
| **Flux-Core (FCAW)** | High deposition | Limited stainless wires | Structural stainless |

## FILLER METAL SELECTION

### TIG Filler Rods (GTAW)

| Base Metal | Filler | Notes |
|---|---|---|
| 304 / 304L | **308L** | Standard choice |
| 316 / 316L | **316L** | Marine grade |
| Unknown stainless | **309L** | Best for dissimilar metals |
| Stainless to carbon steel | **309L** | Prevents cracking |
| 409 / 430 | **409Cb** or **308L** | 308L works for repairs |
| 410 / 420 | **410** or **309L** | 309L reduces cracking |

**MIG Solid Wire (GMAW)**

| Wire | Use Case | Gas | Notes |
|---|---|---|---|
| **ER308L** | 304 stainless | 98/2 Ar-$O_2$ | Most common |
| **ER316L** | 316 stainless | 98/2 | Marine grade |
| **ER309L** | Dissimilar metals | 98/2 | Stainless to carbon |
| **ER409** | Exhaust systems | 98/2 | Automotive |

**Stick Electrodes (SMAW)**

| Rod | Use Case | Notes |
|---|---|---|
| **308L-16** | 304 stainless | Smooth bead |
| **316L-16** | 316 stainless | Marine |
| **309L-16** | Dissimilar metals | Stainless to carbon |
| **410-16** | Martensitic | Preheat required |

## SETTINGS & PARAMETERS

These are practical, real-world ranges.

### TIG (GTAW) SETTINGS — STAINLESS

| Thickness | Tungsten | Amps | Gas Flow | Notes |
|---|---|---|---|---|
| 0.035" | 1/16" | 20–40 | 10–15 CFH | Thin sheet |
| 1/16" | 1/16" | 40–70 | 12–15 CFH | Most common |
| 1/8" | 3/32" | 80–120 | 15–20 CFH | Use 308L/316L |
| 3/16" | 1/8" | 120–180 | 18–22 CFH | Watch heat input |

### MIG (GMAW) SETTINGS — 308L / 316L

**0.030" Wire**

| Thickness | Volts | WFS (ipm) | Notes |
|---|---|---|---|
| 16 ga | 16–17 | 200–260 | Stitch to avoid warp |
| 1/8" | 17–18.5 | 240–300 | Clean metal only |
| 3/16" | 18.5–20 | 300–360 | Good penetration |

**0.035" Wire**

| Thickness | Volts | WFS (ipm) |
| --- | --- | --- |
| 1/8" | 17–19 | 200–280 |
| 3/16" | 19–20.5 | 280–340 |
| 1/4" | 20.5–22 | 340–400 |

**STICK (SMAW) SETTINGS — 308L / 316L / 309L**

| Rod Size | Amps | Notes |
| --- | --- | --- |
| 3/32" | 60–90 | Good for thin stainless |
| 1/8" | 80–110 | Most common |
| 5/32" | 110–150 | Heavy plate |

## WELDING TECHNIQUE NOTES

### TIG

- Keep arc tight
- Use **back purging** for pipe or sanitary welds
- Avoid overheating — stainless warps easily

**MIG**

- Push angle
- Keep metal clean
- Avoid weaving — use straight beads

**Stick**

- Use short arc
- Keep rods dry
- Expect slag; chip thoroughly

## COMMON PROBLEMS & FIXES (STAINLESS)

| Problem | Likely Cause | Fast Fix |
|---|---|---|
| Sugaring (oxidation) | No back purge | Add purge gas, reduce heat |
| Warping | Too much heat | Stitch weld, clamp, reduce amps |
| Porosity | Dirty metal | Clean with acetone, stainless brush |
| Cracking | Wrong filler | Use 309L for unknown metals |
| Discoloration | Overheating | Reduce heat, increase travel |

## CHAPTER 14 — ALUMINUM ALLOYS (1xxx–7xxx SERIES)

*(Sheet aluminum, structural aluminum, cast aluminum, marine alloys, aircraft alloys)*

### IDENTIFICATION

**Common names:**
Aluminum, alloy aluminum, cast aluminum, 6061, 5052, 3003, 4043 filler, 5356 filler.

**How to recognize it:**

- Bright silver, lightweight
- Non-magnetic
- Produces **no sparks** when ground
- Melts suddenly (no color change before collapse)
- Often has oxide layer (white/gray chalky surface)

**Typical uses:**

- Automotive parts
- Boats and marine structures
- Aircraft components
- Frames, brackets, machine parts
- Radiators, intercoolers
- Cast housings and engine parts

**Weldability:** ★★★☆☆ (Good, but requires cleanliness and correct filler)
Aluminum is easy to weld **once clean**, but heat control is critical.

## ALUMINUM SERIES OVERVIEW

### 1xxx Series (Pure Aluminum)

- Very soft
- Excellent corrosion resistance
- Used for electrical and chemical applications
- Welds easily with 1100 filler

### 3xxx Series (3003, 3004)

- Common sheet aluminum
- Good formability
- Welds easily with 4043 filler

### 5xxx Series (5052, 5083, 5086)

- Marine grade
- Stronger than 3xxx
- Welds best with **5356** filler
- Avoid high heat to prevent sensitization

**6xxx Series (6061, 6063)**

- Most common structural aluminum
- Heat-treatable
- Welds well with **4043** or **5356**
- Loses strength in the HAZ

**7xxx Series (7075, 7050)**

- Aircraft grade
- Very strong
- **Not recommended for welding**
- Cracks easily

**Cast Aluminum (A356, A357, misc.)**

- Porous
- Often contaminated
- Welds best with **4043**
- Preheat helps

## PREHEAT & INTERPASS GUIDANCE

**Sheet aluminum (1xxx, 3xxx, 5xxx)**

- Preheat **not required**
- If cold shop: **150–200°F**
- Interpass: **keep below 300°F**

**6061 / 6063 (6xxx series)**

- Preheat **150–250°F**
- Interpass **below 350°F**
- Avoid overheating — reduces strength

**Cast aluminum**

- Preheat **250–350°F**
- Helps drive out moisture and oil
- Interpass **below 350°F**

**7xxx series**

- Avoid welding unless absolutely necessary
- If welding: use **5356**, low heat, small beads

| Material Thickness | Amperage Range (AC) | Tungsten Diameter | Filler Metal (ER4043 / ER5356) | Shielding Gas (100% Argon) | AC Balance (% EN) |
|---|---|---|---|---|---|
| **1/16" (1.6 mm)** | 50 – 80 A | 3/32" (2.4 mm) | 1/16" (1.6 mm) | 15 – 20 CFH | 65% – 70% |
| **1/8" (3.2 mm)** | 125 – 160 A | 3/32" – 1/8" | 3/32" (2.4 mm) | 20 CFH | 65% – 75% |
| **1/4" (6.4 mm)** | 180 – 225 A | 1/8" (3.2 mm) | 1/8" (3.2 mm) | 20 – 25 CFH | 70% – 75% |
| **3/8" (9.5 mm)+** | 230 – 300 A+ *(Preheat Required)* | 5/32" (4.0 mm) | 5/32" (4.0 mm) | 25 – 30 CFH *(or Ar/He Mix)* | 70% – 75% |

## BEST PROCESSES FOR ALUMINUM

| Process | Strengths | Weaknesses | Best Use Cases |
|---|---|---|---|
| **TIG (GTAW)** | Cleanest, precise | Slow | Thin sheet, cast aluminum |
| **MIG (GMAW)** | Fast, strong | Needs spool gun or push-pull | Structural aluminum |
| **Stick (SMAW)** | Rarely used | Poor control | Emergency repairs only |
| **Flux-Core** | Not used | — | Aluminum FCAW is specialty only |

## FILLER METAL SELECTION

### TIG & MIG Filler Rods/Wire

| Base Metal | Filler | Notes |
|---|---|---|
| 3003 | **4043** | Best match |
| 5052 | **5356** | Marine grade |

| Base Metal | Filler | Notes |
|---|---|---|
| 5083 / 5086 | **5356** | High strength |
| 6061 | **4043** (best) / **5356** (stronger) | 4043 reduces cracking |
| Cast aluminum | **4043** | Flows better |
| 7xxx | **5356** | Only for emergency repairs |

**4043** = best flow, least cracking
**5356** = stronger, stiffer, better for marine

### SETTINGS & PARAMETERS

These are practical field ranges.

### TIG (GTAW) SETTINGS — ALUMINUM (AC)

| Thickness | Tungsten | Amps | Gas Flow | Notes |
|---|---|---|---|---|
| 0.040" | 1/16" | 30–50 | 12–15 CFH | Thin sheet |
| 1/16" | 1/16" | 50–80 | 12–15 CFH | Most common |
| 1/8" | 3/32" | 90–140 | 15–20 CFH | 4043 or 5356 |
| 3/16" | 1/8" | 140–200 | 18–22 CFH | Preheat helps |
| 1/4" | 1/8" | 200–260 | 20–25 CFH | Heavy plate |

**AC balance:**

- Start at **70% EN**
- More cleaning = increase EP
- More penetration = increase EN

**MIG (GMAW) SETTINGS — ALUMINUM (4043 or 5356)**

**0.030" Wire (Spool Gun)**

| Thickness | Volts | WFS (ipm) | Notes |
|---|---|---|---|
| 1/16" | 17–18 | 350–450 | Fast travel |
| 1/8" | 18–20 | 450–550 | Most common |
| 3/16" | 20–22 | 550–650 | Good penetration |

**0.035" Wire**

| Thickness | Volts | WFS (ipm) |
|---|---|---|
| 1/8" | 18–20 | 350–450 |
| 3/16" | 20–22 | 450–550 |
| 1/4" | 22–24 | 550–650 |

**WELDING TECHNIQUE NOTES**

**TIG**

- Clean metal with stainless brush
- Use AC balance for oxide cleaning
- Keep arc tight
- Add filler quickly to avoid collapse

**MIG**

- Use **push angle only**
- Keep stickout short
- Use spool gun or push-pull system
- Travel fast — aluminum melts suddenly

**Cast aluminum**

- Preheat
- Expect porosity
- Use 4043
- Slow cool

**COMMON PROBLEMS & FIXES (ALUMINUM)**

| **Problem** | **Likely Cause** | **Fast Fix** |
|---|---|---|
| Porosity | Oil, moisture, dirty metal | Clean thoroughly, preheat |
| Black soot | Wrong polarity or gas | Use pure argon, check AC balance |
| Collapsing puddle | Too slow, too hot | Move faster, reduce heat |
| Cracking | Wrong filler | Use 4043 for most jobs |
| Lack of fusion | Too cold | Increase amps/volts |

## CHAPTER 15 — CAST IRON

*(Gray iron, ductile iron, malleable iron, engine blocks, housings, manifolds)*

### IDENTIFICATION

**Common names:**
Cast iron, gray iron, ductile iron, nodular iron, malleable iron, engine block iron.

**How to recognize it:**

- Dark gray, grainy appearance
- Very brittle — breaks rather than bends
- Produces **short, red-orange sparks** (much shorter than steel)
- Often porous or oil-soaked
- Rings with a dull thud when struck
- Heavy for its size

**Typical uses:**

- Engine blocks
- Gear housings
- Pump bodies
- Exhaust manifolds
- Vises, anvils
- Machine bases
- Wood-burning stoves

**Weldability:** ★★☆☆☆ (Challenging — requires correct filler and heat control)
Cast iron cracks easily if welded incorrectly.

**TYPES OF CAST IRON**

**Gray Iron (Most common)**

- Flake graphite structure
- Brittle
- Easy to crack
- Weldable with **nickel rods** or **cold-weld technique**

**Ductile Iron (Nodular Iron)**

- Spherical graphite
- Stronger and more ductile
- Weldable with **nickel** or **steel fillers**
- Better for structural repairs

**Malleable Iron**

- Heat-treated cast iron
- Stronger than gray iron
- Weldable but heat-sensitive
- Often brazed instead of welded

**White Iron**

- Extremely hard
- **Not weldable**
- Will crack immediately
- Must be replaced or buttered with nickel and machined

**PREHEAT & INTERPASS GUIDANCE**

Cast iron welding is all about **controlling expansion and contraction**.

**Hot-Weld Method (Traditional)**

- Preheat entire part to **500–1200°F**
- Maintain heat during welding
- Slow cool in sand, ashes, or insulation
- Best for large, thick parts (engine blocks, housings)

**Cold-Weld Method (Most common in field)**

- **No preheat** or minimal preheat (100–200°F)
- Very short welds (1/2" max)
- Peen each bead immediately
- Allow cooling between passes
- Best for small cracks, thin castings

**Ductile Iron**

- Preheat **300–600°F**
- Slow cool
- Use nickel or steel filler depending on strength needed

**Malleable Iron**

- Preheat **400–600°F**
- Slow cool
- Avoid overheating — can revert to brittle white iron

## BEST PROCESSES FOR CAST IRON

| Process | Strengths | Weaknesses | Best Use Cases |
|---|---|---|---|
| **Stick (SMAW)** | Best control, nickel rods | Slow, expensive rods | Engine blocks, housings |
| **TIG (GTAW)** | Clean, precise | Requires clean casting | Small cracks, thin cast |
| **MIG (GMAW)** | Fast | High cracking risk | Only with nickel wire |
| **Brazing (Oxy-fuel or TIG)** | Low heat, flexible | Not as strong | Manifolds, thin castings |

## FILLER METAL SELECTION

### Stick Electrodes (SMAW)

| Rod | Use Case | Notes |
|---|---|---|
| **Ni-99 (ENi-CI)** | Best for gray iron | Soft, machinable |
| **Ni-55 (ENiFe-CI)** | Stronger, less machinable | Good for ductile iron |
| **7018** | Ductile iron only | Not for gray iron |
| **Cast iron repair rods** | General repairs | Manufacturer-specific |

**Recommended all-around rod: Ni-99**

### TIG Filler Rods (GTAW)

| Filler | Use Case | Notes |
|---|---|---|
| **Nickel 99 TIG rod** | Gray iron | Best for crack resistance |
| **Nickel 55 TIG rod** | Ductile iron | Stronger |
| **Silicon bronze (ERCuSi-A)** | Brazing | Low heat, flexible |

**MIG Wire (GMAW)**

*(Used rarely — only for specialty repairs)*

| Wire | Use Case | Notes |
| --- | --- | --- |
| **Ni-99 MIG wire** | Gray iron | Expensive, specialty |
| **Ni-55 MIG wire** | Ductile iron | Stronger |

## SETTINGS & PARAMETERS

These are practical field ranges.

### STICK (SMAW) SETTINGS — CAST IRON

**Ni-99 / Ni-55**

| Rod Size | Amps | Notes |
| --- | --- | --- |
| 3/32" | 60–80 | Best for cold-weld |
| 1/8" | 80–100 | Short beads only |
| 5/32" | 100–130 | Hot-weld method |

## TIG (GTAW) SETTINGS — CAST IRON

| Thickness | Tungsten | Amps | Gas Flow | Notes |
|---|---|---|---|---|
| 1/8" | 3/32" | 70–110 | 12–18 CFH | Nickel rod |
| 3/16" | 1/8" | 110–150 | 15–20 CFH | Preheat helps |
| 1/4" | 1/8" | 150–200 | 18–22 CFH | Slow cool |

## BRAZING SETTINGS (TIG Brazing)

| Filler | Amps | Notes |
|---|---|---|
| Silicon bronze | 40–120 | Low heat, great for manifolds |

## WELDING TECHNIQUE NOTES

### Cold-Weld Method (Most common)

- Weld **1/2" at a time**
- Peen immediately
- Let cool completely
- Repeat
- Prevents cracking by relieving stress

### Hot-Weld Method

- Heat entire part evenly
- Weld normally
- Slow cool in sand or insulation
- Best for large, thick castings

### Brazing

- Excellent for exhaust manifolds
- Low heat = less cracking
- Not as strong as welding but very reliable

## COMMON PROBLEMS & FIXES (CAST IRON)

| Problem | Likely Cause | Fast Fix |
|---|---|---|
| Cracking | Too much heat | Use cold-weld method |
| Porosity | Oil contamination | Preheat, clean thoroughly |
| Hard spots | Overheating | Reduce heat, use nickel |
| Weld won't stick | Oxides | Grind clean, preheat |
| Brittle weld | Wrong filler | Use Ni-99 or Ni-55 |

## CHAPTER 16 — GALVANIZED STEEL

*(Hot-dip galvanized, electro-galvanized, zinc-coated sheet, farm steel, structural components)*

### IDENTIFICATION

**Common names:**
Galvanized steel, zinc-coated steel, hot-dip galvanized, electro-galvanized, GI sheet.

**How to recognize it:**

- Silver, spangled, crystal-patterned surface (hot-dip)
- Smooth, dull gray surface (electro-galvanized)
- Produces **white smoke** when welded (zinc fumes)
- Magnetic (base metal is carbon steel)
- Often used outdoors or anywhere corrosion resistance is needed

**Typical uses:**

- Farm gates, fencing, posts
- HVAC ducting
- Trailers
- Guardrails
- Roofing and siding panels
- Structural components exposed to weather

**Weldability:** ★★★☆☆ (Good, but requires safety precautions and prep)
Galvanized steel welds like mild steel **once the zinc is removed**.

## SAFETY WARNING — ZINC FUMES

Welding galvanized steel produces **zinc oxide fumes**, which can cause:

- Metal fume fever
- Nausea
- Headache
- Chills
- Flu-like symptoms

**Always:**

- Grind off zinc coating at least **1–2 inches** from weld area
- Use **excellent ventilation**
- Use a **respirator** when needed
- Avoid welding in enclosed spaces

## PREHEAT & INTERPASS GUIDANCE

Galvanized steel behaves like mild steel once cleaned.

**General guidance:**

- **No preheat required** for thin material
- For thicker sections (> 1/4"): preheat **100–150°F**
- Interpass temperature: **below 400°F**

**Notes:**

- Removing zinc reduces porosity
- Excessive heat burns zinc and increases fumes
- Clean metal = cleaner welds

## BEST PROCESSES FOR GALVANIZED STEEL

| Process | Strengths | Weaknesses | Best Use Cases |
|---|---|---|---|
| **Stick (SMAW)** | Works on dirty steel | More fumes | Farm repairs, outdoor |
| **MIG (GMAW)** | Clean, fast | Needs clean metal | Fabrication, shop work |
| **Flux-Core (FCAW)** | Good for outdoor | More fumes, slag | Field repairs |
| **TIG (GTAW)** | Cleanest | Requires full zinc removal | Thin sheet, precision |

## FILLER METAL SELECTION

Galvanized steel uses the same fillers as **mild steel.**

### Stick Electrodes (SMAW)

| Rod | Use Case | Notes |
|---|---|---|
| **6011** | Dirty/galvanized steel | Best for burning through zinc |
| **6013** | Thin galvanized sheet | Smooth bead |
| **7018** | Cleaned galvanized | Strongest weld |

**Recommended all-around rod: 6011** for dirty steel, **7018** for cleaned steel.

•

### MIG Solid Wire (GMAW)

| Wire | Gas | Notes |
|---|---|---|
| **ER70S-6** | 75/25 | Best for galvanized |
| **ER70S-3** | 75/25 | For clean steel only |

### Flux-Core Wire (FCAW)

| Wire | Shielding | Notes |
|---|---|---|
| **E71T-11** | Self-shielded | Great for outdoor galvanized |
| **E71T-1** | Gas-shielded | Cleaner bead |

**TIG Filler Rods (GTAW)**

| Filler | Notes |
|---|---|
| ER70S-2 | Best for TIG on galvanized |
| ER70S-6 | Works if metal is clean |

## SETTINGS & PARAMETERS

These are practical field ranges.

## STICK (SMAW) SETTINGS — GALVANIZED

**6011**

| Rod Size | Amps | Notes |
|---|---|---|
| 3/32" | 60–90 | Good for thin galvanized |
| 1/8" | 85–120 | Burns through zinc well |

**7018**

| Rod Size | Amps | Notes |
|---|---|---|
| 1/8" | 115–135 | Only after cleaning zinc |

## MIG (GMAW) SETTINGS — ER70S-6

### 0.030" Wire

| Thickness | Volts | WFS (ipm) | Notes |
|---|---|---|---|
| 16 ga | 16–17 | 200–260 | Stitch to avoid burn-through |
| 1/8" | 17–18.5 | 260–320 | Clean zinc first |
| 3/16" | 18.5–20 | 320–380 | Good penetration |

## FLUX-CORE (FCAW) SETTINGS — E71T-11

### 0.035" Wire

| Thickness | Volts | WFS (ipm) | Notes |
|---|---|---|---|
| 1/8" | 18–19 | 250–300 | Great for outdoor |
| 3/16" | 19–20 | 300–350 | Burns through zinc |
| 1/4" | 20–21 | 350–400 | Multi-pass |

## TIG (GTAW) SETTINGS — GALVANIZED

| Thickness | Tungsten | Amps | Gas Flow | Notes |
|---|---|---|---|---|
| 1/16" | 1/16" | 40–70 | 10–15 CFH | Remove zinc completely |
| 1/8" | 3/32" | 80–120 | 12–18 CFH | Clean metal only |

## WELDING TECHNIQUE NOTES

### General

- Grind off zinc coating **1–2 inches** around weld
- Ventilation is critical
- Expect more spatter and fumes

### Stick

- 6011 is excellent for galvanized
- Use whip/pause technique

### MIG

- Push angle
- Clean metal thoroughly
- Expect some porosity if zinc remains

### Flux-Core

- Drag angle
- Works well outdoors
- More fumes — ventilate

### TIG

- Must remove all zinc
- Use ER70S-2
- Keep heat low

**COMMON PROBLEMS & FIXES (GALVANIZED)**

| Problem | Likely Cause | Fast Fix |
|---|---|---|
| Porosity | Zinc contamination | Grind zinc off |
| Excessive spatter | Zinc burning | Clean metal, reduce heat |
| White smoke | Zinc fumes | Ventilate, respirator |
| Poor bead appearance | Zinc left on metal | Clean thoroughly |
| Cracking | Overheating | Reduce heat, use 6011 |

# CHAPTER 17 COPPER & COPPER ALLOYS

*(Copper, brass, bronze, silicon bronze, beryllium copper, plumbing tube, electrical bus bars)*

## IDENTIFICATION

**Common names:**
Copper, brass, bronze, silicon bronze, phosphor bronze, beryllium copper, plumbing copper, electrical copper.

**How to recognize it:**

- **Copper:** Reddish-orange, soft, very conductive
- **Brass:** Yellow-gold color (copper + zinc)
- **Bronze:** Brownish-gold (copper + tin)
- **Silicon bronze:** Golden-brown, used for brazing
- **Beryllium copper:** Looks like bronze but harder
- All produce **long, red sparks** when ground (but less than steel)
- Non-magnetic

**Typical uses:**

- Plumbing lines
- Electrical bus bars
- Heat exchangers
- Bearings and bushings
- Marine hardware
- Decorative metalwork
- Brazing filler for steel

**Weldability:** ★★★☆☆ (Good with TIG; tricky with MIG; brazing is common)
Copper pulls heat away fast — requires **high amps** and **clean metal**.

**COPPER ALLOY TYPES**

**Pure Copper (C110, C122)**

- Very soft
- Extremely conductive
- Used for electrical and plumbing
- Weldable with TIG or brazing
- Requires high heat input

**Brass (Copper + Zinc)**

- Yellow color
- Melts suddenly
- Zinc fumes are hazardous
- Best joined by **brazing**, not welding

**Bronze (Copper + Tin)**

- Harder than brass
- Good wear resistance
- Weldable with silicon bronze

**Silicon Bronze (ERCuSi-A)**

- Filler metal AND base metal
- Excellent for TIG brazing
- Low heat input
- Great for sheet metal and dissimilar metals

**Beryllium Copper**

- Very strong
- Used for springs, tools
- **Do not weld** — toxic fumes
- Brazing or mechanical fastening only

**PREHEAT & INTERPASS GUIDANCE**

Copper alloys conduct heat extremely fast.

**Pure Copper**

- Preheat **300–600°F**
- Interpass **below 700°F**
- Helps puddle form faster

**Brass**

- Preheat **not required**
- Keep heat low
- Avoid zinc fumes

**Bronze**

- Preheat **200–400°F**
- Slow cool

**Silicon Bronze (as base metal)**

- Preheat **not required**
- Low heat input
- Excellent for thin sheet

## BEST PROCESSES FOR COPPER & ALLOYS

| Process | Strengths | Weaknesses | Best Use Cases |
|---|---|---|---|
| **TIG (GTAW)** | Best control, clean | High heat needed | Copper, bronze, silicon bronze |
| **Brazing (TIG or oxy-fuel)** | Low heat, flexible | Not as strong | Brass, bronze, sheet metal |
| **MIG (GMAW)** | Rarely used | Hard to control | Specialty only |
| **Stick (SMAW)** | Not common | Poor control | Emergency repairs |

## FILLER METAL SELECTION

### TIG Filler Rods (GTAW)

| Base Metal | Filler | Notes |
|---|---|---|
| Pure copper | **ERCu** | Best match |
| Bronze | **ERCuSn-A** | Tin bronze |
| Silicon bronze | **ERCuSi-A** | Most versatile |
| Brass | **Silicon bronze** | Avoid welding brass directly |
| Copper to steel | **Silicon bronze** | Excellent for dissimilar metals |

**Silicon bronze (ERCuSi-A)** is the **most useful** filler in this entire chapter.

### Brazing Rods

| Rod | Use Case | Notes |
|---|---|---|
| **Silicon bronze** | Steel, copper, brass | Low heat, strong |
| **Phosphor bronze** | Copper plumbing | Good flow |
| **Brass brazing rod** | General repairs | Use with caution (zinc fumes) |

## SETTINGS & PARAMETERS

These are practical field ranges.

## TIG (GTAW) SETTINGS — COPPER & ALLOYS

### Pure Copper

| Thickness | Tungsten | Amps | Gas Flow | Notes |
|---|---|---|---|---|
| 1/16" | 1/16" | 80–120 | 12–18 CFH | Preheat helps |
| 1/8" | 3/32" | 120–200 | 15–20 CFH | High heat needed |
| 3/16" | 1/8" | 200–260 | 18–22 CFH | Very conductive |

### Silicon Bronze (TIG Brazing)

| Thickness | Amps | Notes |
|---|---|---|
| Sheet metal | 40–80 | Low heat, fast travel |
| 1/8" | 80–120 | Great for dissimilar metals |

## BRAZING (Oxy-fuel)

| Rod | Heat | Notes |
|---|---|---|
| Silicon bronze | Medium | Clean, strong |
| Brass rod | Medium-high | Watch zinc fumes |

## WELDING TECHNIQUE NOTES

### Pure Copper

- Preheat is your friend
- Copper pulls heat away fast
- Use high amps
- Keep arc tight

### Brass

- Avoid welding — **braze instead**
- Zinc fumes are hazardous
- Use silicon bronze

### Bronze

- Preheat helps
- Use ERCuSn-A
- Slow cool

### Silicon Bronze (TIG Brazing)

- Low heat
- Great for sheet metal
- Excellent for copper-to-steel
- Produces beautiful golden beads

### COMMON PROBLEMS & FIXES (COPPER & ALLOYS)

| Problem | Likely Cause | Fast Fix |
| --- | --- | --- |
| Puddle won't form | Copper pulling heat | Preheat, increase amps |
| Porosity | Contamination | Clean thoroughly |
| Cracking | Wrong filler | Use silicon bronze |
| Zinc fumes | Welding brass | Braze instead |
| Overheating | Too much heat | Reduce amps, faster travel |

# CHAPTER 18 — TOOL STEELS (OVERVIEW & WARNINGS)

*(O-series, W-series, A-series, D-series, H-series, S-series, high-speed steels)*

## IDENTIFICATION

**Common names:**
Tool steel, hardened steel, O1, W1, A2, D2, H13, S7, high-speed steel (HSS), die steel, punch steel.

**How to recognize it:**

- Very hard surface
- Often used for cutting, shaping, or impact tools
- Produces **bright, dense sparks** when ground
- Rings sharply when struck
- Often shows wear patterns instead of bending
- May be oil-stained from heat-treating

**Typical uses:**

- Punches and dies
- Shear blades
- Knives and cutting tools
- Molds
- Shafts and spindles
- Hammers and chisels
- High-temperature tooling (H13)

**Weldability:** ★☆☆☆☆ (Poor — extremely heat-sensitive)
Tool steels can **crack, warp, or lose hardness** if welded incorrectly.

**TOOL STEEL GROUPS**

**W-Series (Water-Hardening)**

- Very hard
- Very brittle
- **Worst for welding**
- Cracks easily
- Avoid welding unless absolutely necessary

**O-Series (Oil-Hardening)**

- Common in machine shops
- Moderate weldability
- Requires **preheat and slow cool**

**A-Series (Air-Hardening)**

- High wear resistance
- Very crack-sensitive
- Requires **high preheat** and controlled cooling

**D-Series (High Carbon, High Chromium)**

- Extremely wear-resistant
- Very hard
- **Difficult to weld**
- Often repaired by **brazing** instead

**H-Series (Hot-Work Steels: H11, H13)**

- Used for dies, molds, high-heat tooling
- Better weldability than most tool steels
- Requires **strict heat control**

**S-Series (Shock-Resistant: S7)**

- Used for chisels, punches, hammers
- Good impact resistance
- Weldable with correct preheat

**High-Speed Steels (M2, M42, T1)**

- Used for drill bits, end mills, cutting tools
- **Not weldable** — will crack or lose hardness
- Replace or braze only

## PREHEAT & INTERPASS GUIDANCE

Tool steels require **strict heat control.**

### General Rules

- Preheat is **mandatory** for almost all tool steels
- Interpass temperature must be controlled
- Slow cooling is essential
- Post-heat often required

### Typical Preheat Ranges

| Tool Steel Type | Preheat | Interpass Max | Notes |
|---|---|---|---|
| W-Series | **Do not weld** | — | Replace or braze |
| O-Series | 400–600°F | 600°F | Slow cool |
| A-Series | 800–1000°F | 1000°F | Very crack-sensitive |
| D-Series | 800–1000°F | 1000°F | Often brazed instead |
| H-Series (H13) | 600–900°F | 900°F | Best tool steel to weld |
| S-Series (S7) | 400–600°F | 600°F | Good impact steel |
| High-Speed Steel | **Do not weld** | — | Replace or braze |

## BEST PROCESSES FOR TOOL STEELS

| Process | Strengths | Weaknesses | Best Use Cases |
|---|---|---|---|
| **TIG (GTAW)** | Best control | Slow, requires skill | Small repairs, cracks |
| **Stick (SMAW)** | Good penetration | Hard to control heat | Larger repairs |
| **MIG (GMAW)** | Rarely used | Too much heat | Not recommended |
| **Brazing (TIG or oxy-fuel)** | Low heat | Not as strong | D-series, W-series |

## FILLER METAL SELECTION

Tool steel welding is usually done with **nickel**, **stainless**, or **tool-steel-specific fillers**.

### TIG Filler Rods (GTAW)

| Base Metal | Filler | Notes |
|---|---|---|
| O-Series | **ER312** or **Ni-99** | Crack-resistant |
| A-Series | **ER312** | Best for air-hardening steels |
| D-Series | **ER312** or **brazing** | Welding often avoided |
| H-Series (H13) | **H13 filler** | Best match |
| S-Series (S7) | **ER312** | Good impact strength |

| Base Metal | Filler | Notes |
|---|---|---|
| Unknown tool steel | **ER312** | Universal tool-steel filler |

**ER312** is the **go-to filler** for unknown or mixed tool steels.

### Stick Electrodes (SMAW)

| Rod | Use Case | Notes |
|---|---|---|
| **312 stainless rod** | Most tool steels | Crack-resistant |
| **Nickel 99** | O-series, unknown steels | Soft, machinable |
| **H13 rod** | H-series | Strength-matched |

### Brazing Rods

| Rod | Use Case | Notes |
|---|---|---|
| **Silicon bronze** | D-series, W-series | Low heat, flexible |
| **Silver braze** | Precision tools | Very clean joints |

## SETTINGS & PARAMETERS

These are practical field ranges.

## TIG (GTAW) SETTINGS — TOOL STEELS

| Thickness | Tungsten | Amps | Gas Flow | Notes |
|---|---|---|---|---|
| 1/8" | 3/32" | 80–120 | 12–18 CFH | Preheat required |
| 3/16" | 1/8" | 120–180 | 15–20 CFH | Use ER312 |
| 1/4" | 1/8" | 150–220 | 18–22 CFH | Slow cool |

## STICK (SMAW) SETTINGS — TOOL STEELS

### 312 Stainless Rod

| Rod Size | Amps | Notes |
|---|---|---|
| 3/32" | 70–90 | Good for cracks |
| 1/8" | 90–120 | Most common |
| 5/32" | 120–160 | Heavy repairs |

## WELDING TECHNIQUE NOTES

### General

- Preheat is mandatory
- Use **stringer beads only**
- Avoid weaving
- Slow cool in sand, insulation, or furnace
- Peen lightly to relieve stress

**TIG**

- Tight arc
- Use ER312 unless you know the exact alloy
- Control heat carefully

**Stick**

- Short arc
- Use 312 stainless rods
- Maintain preheat

**Brazing**

- Best for D-series and W-series
- Low heat = less cracking
- Great for precision tools

## COMMON PROBLEMS & FIXES (TOOL STEELS)

| Problem | Likely Cause | Fast Fix |
|---|---|---|
| Cracking | No preheat | Preheat 400–1000°F |
| Brittle weld | Wrong filler | Use ER312 |
| Loss of hardness | Overheating | Reduce heat, slow cool |
| Porosity | Contamination | Clean thoroughly |
| Warping | Uneven heat | Preheat entire part |

## SECTION 3 — FILLER METALS & CONSUMABLES

## CHAPTER 19 — STICK ELECTRODES (SMAW)

*(6010, 6011, 6013, 7014, 7018, 7024 — what they do, where they shine, and how to run them)*

### UNDERSTANDING ROD CLASSIFICATIONS

A stick electrode classification (like **7018**) tells you everything you need to know:

- **First two digits:** Tensile strength
  - 60 = 60,000 psi
  - 70 = 70,000 psi
- **Third digit:** Welding position
  - 1 = All positions
  - 2 = Flat & horizontal only
- **Fourth digit:** Flux type & polarity
  - 0,1 = Cellulosic
  - 3 = Rutile
  - 4 = Iron powder rutile
  - 8 = Low-hydrogen

**ELECTRODE OVERVIEW TABLE**

| Rod | Strength | Positions | Polarity | Best For |
|---|---|---|---|---|
| **6010** | 60 ksi | All | DCEP | Pipe roots, dirty steel, deep penetration |
| **6011** | 60 ksi | All | AC / DCEP | Farm repairs, rust, paint, poor fit-up |
| **6013** | 60 ksi | All | AC / DCEP | Thin sheet, smooth bead, low penetration |
| **7014** | 70 ksi | Flat/Horizontal | AC / DCEP | Fast fill, smooth bead, iron powder |
| **7018** | 70 ksi | All | DCEP | Structural, code work, clean steel |
| **7024** | 70 ksi | Flat/Horizontal | AC / DCEP | High deposition, “drag rod,” production |

## 6010 — THE DEEP-PENETRATION ROOT ROD

**Type:** Cellulosic
**Polarity:** DCEP only
**Strength:** 60,000 psi
**Positions:** All

**What it's best for**

- Pipe roots
- Dirty, rusty, painted steel
- Poor fit-up
- Fast-freeze applications
- Open-root passes

**Technique**

- Whip and pause
- Keep arc long enough to dig
- Fast travel

**Settings (Practical Field Ranges)**

| Rod Size | Amps |
|---|---|
| 3/32" | 70–90 |
| 1/8" | 85–120 |
| 5/32" | 110–160 |

### 6011 — THE FARM & FIELD ROD

**Type:** Cellulosic
**Polarity:** AC / DCEP
**Strength:** 60,000 psi
**Positions:** All

**What it's best for**

- Outdoor repairs
- Rusty steel
- Galvanized steel
- AC buzz box machines
- Farm equipment

**Technique**

- Whip/pause like 6010
- Works great on AC machines

**Settings**

| Rod Size | Amps |
|---|---|
| 3/32" | 60–90 |
| 1/8" | 85–120 |
| 5/32" | 110–160 |

### 6013 — THE SHEET METAL ROD

**Type:** Rutile
**Polarity:** AC / DCEP

**Strength:** 60,000 psi
**Positions:** All

**What it's best for**

- Thin sheet
- Auto body
- Smooth, cosmetic welds
- Low penetration jobs

**Technique**

- Short arc
- Slow travel
- Minimal penetration

**Settings**

| Rod Size | Amps |
| --- | --- |
| 3/32" | 40–70 |
| 1/8" | 80–110 |

## 7014 — THE FAST-FILL IRON POWDER ROD

**Type:** Iron powder rutile
**Polarity:** AC / DCEP
**Strength:** 70,000 psi
**Positions:** Flat & horizontal

**What it's best for**

- Flat fillet welds
- Production work
- Smooth, wide beads
- Filling large gaps

**Technique**

- Drag rod
- High deposition
- Very smooth bead appearance

**Settings**

| Rod Size | Amps |
|---|---|
| 1/8" | 110–150 |
| 5/32" | 150–200 |

## 7018 — THE LOW-HYDROGEN STRUCTURAL ROD

**Type:** Low-hydrogen
**Polarity:** DCEP
**Strength:** 70,000 psi
**Positions:** All

**What it's best for**

- Structural steel
- Code work
- High-strength applications
- Clean steel
- Multi-pass welds

**Technique**

- Short arc
- Drag or slight push
- Keep rods dry (rod oven recommended)

**Settings**

| Rod Size | Amps |
|---|---|
| 3/32" | 70–100 |
| 1/8" | 115–135 |
| 5/32" | 150–190 |

## 7024 — THE "DRAG ROD" FOR PRODUCTION

**Type:** Iron powder high-fill
**Polarity:** AC / DCEP
**Strength:** 70,000 psi
**Positions:** Flat & horizontal

### What it's best for

- Heavy fillet welds
- Long flat beads
- High-speed production
- Thick plate

### Technique

- Drag only
- Very high deposition
- Smooth, wide bead

### Settings

| Rod Size | Amps |
|---|---|
| 1/8" | 140–180 |
| 5/32" | 180–240 |

## ROD SELECTION QUICK CHART

| Job Type | Best Rod | Why |
|---|---|---|
| Dirty/rusty steel | **6011** | Burns through contamination |
| Pipe root | **6010** | Deep penetration, fast freeze |
| Structural steel | **7018** | Low-hydrogen, strong |
| Thin sheet | **6013** | Low penetration |
| Flat fillets | **7024** | High deposition |
| General repairs | **6011** | Works on anything |

## COMMON PROBLEMS & FIXES

| Problem | Cause | Fix |
|---|---|---|
| Porosity | Dirty metal | Clean, increase heat |
| Undercut | Too hot, too fast | Reduce amps, slow travel |
| Sticking rod | Too cold | Increase amps |
| Spatter | Long arc | Shorten arc length |
| Cracking | Wrong rod | Use 7018 or low-hydrogen |

## CHAPTER 20 — MIG SOLID WIRE (GMAW)

*(ER70S-3, ER70S-6, stainless wires, aluminum wires — selection, settings, and best uses)*

### UNDERSTANDING MIG WIRE CLASSIFICATIONS

A MIG wire classification (like **ER70S-6**) tells you:

- **ER** = Electrode, Rod
- **70** = Tensile strength (70,000 psi)
- **S** = Solid wire
- **6** = Chemical composition (deoxidizers, alloy content)

**Higher last digits = more deoxidizers**, which means better performance on mill scale or slightly dirty steel.

### SOLID WIRE OVERVIEW TABLE

| Wire | Material | Gas | Best For |
|---|---|---|---|
| **ER70S-3** | Mild steel | 75/25 | Clean steel, sheet metal |
| **ER70S-6** | Mild steel | 75/25 or $CO_2$ | General fabrication, mill scale |
| **ER308L** | Stainless | 98/2 | 304 stainless |
| **ER316L** | Stainless | 98/2 | Marine stainless |
| **ER309L** | Stainless | 98/2 | Stainless to carbon steel |
| **4043** | Aluminum | Argon | 3003, 6061, cast aluminum |

| Wire | Material | Gas | Best For |
|---|---|---|---|
| **5356** | Aluminum | Argon | Marine aluminum, 5xxx series |

**ER70S-3 — CLEAN STEEL WIRE**

**Material:** Mild steel
**Gas:** 75/25
**Strength:** 70,000 psi

**Best for**

- Clean steel
- Sheet metal
- Thin material
- Auto body work

**Notes**

- Lower deoxidizers
- Produces cleaner, smoother beads
- Not ideal for rusty or mill-scale steel

## ER70S-6 — THE ALL-AROUND MILD STEEL WIRE

**Material:** Mild steel
**Gas:** 75/25 or $CO_2$
**Strength:** 70,000 psi

**Best for**

- General fabrication
- Slightly dirty steel
- Mill scale
- Farm repairs
- Structural fabrication

**Notes**

- Most common MIG wire in the world
- Higher deoxidizers = better puddle on imperfect steel
- Works with $CO_2$ for deeper penetration

## STAINLESS MIG WIRES

### ER308L — For 304 Stainless

**Gas:** 98/2 Ar-$O_2$
**Best for:** Food-grade stainless, general stainless fab
**Notes:** Most common stainless wire

### ER316L — For 316 Stainless

**Gas:** 98/2
**Best for:** Marine stainless, chemical environments
**Notes:** Better corrosion resistance

### ER309L — Dissimilar Metals

**Gas:** 98/2
**Best for:**

- Stainless to carbon steel
- Unknown stainless
- Buffer layers

**Notes:** The “problem solver” stainless wire

## ALUMINUM MIG WIRES

### 4043 — The Easy-Flow Aluminum Wire

**Gas:** 100% argon
**Best for:**

- 3003
- 6061
- Cast aluminum

**Notes:**

- Flows beautifully
- Less cracking
- Best all-around aluminum wire

### 5356 — The Strong Aluminum Wire

**Gas:** 100% argon
**Best for:**

- 5052
- 5083
- 5086
- Marine aluminum

**Notes:**

- Stronger than 4043
- Stiffer wire (feeds better)
- Not ideal for anodized aluminum color match

## SETTINGS & PARAMETERS

These are **real-world, practical ranges** welders actually use.

### MILD STEEL — ER70S-6 SETTINGS

#### 0.023" Wire (Thin Sheet)

| Thickness | Volts | WFS (ipm) | Notes |
|---|---|---|---|
| 22–20 ga | 14–16 | 180–240 | Auto body |
| 1/16" | 15–16.5 | 200–260 | Stitch weld |

#### 0.030" Wire (Most Common)

| Thickness | Volts | WFS (ipm) | Notes |
|---|---|---|---|
| 16 ga | 16–17 | 200–260 | Avoid burn-through |
| 1/8" | 17–18.5 | 260–320 | General fab |
| 3/16" | 18.5–20 | 320–380 | Good penetration |
| 1/4" | 20–21 | 380–450 | Multi-pass |

#### 0.035" Wire

| Thickness | Volts | WFS (ipm) |
|---|---|---|
| 1/8" | 17–19 | 200–280 |
| 3/16" | 19–20.5 | 280–340 |
| 1/4" | 20.5–22 | 340–400 |

## STAINLESS — 308L / 316L / 309L SETTINGS

### 0.030" Wire

| Thickness | Volts | WFS (ipm) | Notes |
|---|---|---|---|
| 16 ga | 16–17 | 200–260 | Clean metal only |
| 1/8" | 17–18.5 | 240–300 | Food-grade fab |
| 3/16" | 18.5–20 | 300–360 | Good penetration |

## ALUMINUM — 4043 / 5356 SETTINGS

### 0.030" Wire (Spool Gun)

| Thickness | Volts | WFS (ipm) | Notes |
|---|---|---|---|
| 1/16" | 17–18 | 350–450 | Fast travel |
| 1/8" | 18–20 | 450–550 | Most common |
| 3/16" | 20–22 | 550–650 | Good penetration |

### 0.035" Wire

| Thickness | Volts | WFS (ipm) |
|---|---|---|
| 1/8" | 18–20 | 350–450 |
| 3/16" | 20–22 | 450–550 |
| 1/4" | 22–24 | 550–650 |

## TECHNIQUE NOTES

### General MIG

- Push angle for cleaner bead
- Keep stickout 3/8"
- Clean metal = better welds

### Stainless MIG

- Push only
- Keep heat low to avoid warping
- Use 98/2 gas

### Aluminum MIG

- Use spool gun or push-pull
- Push angle only
- Travel fast
- Keep stickout short

### FIELD NOTE: SHIELDING GAS SELECTION FOR STAINLESS GMAW

While a **98% Ar / 2% $O_2$** mixture is highly effective for high-deposition **spray transfer** on thicker stainless steel plate, it performs poorly on thin sheet metal. Running 98/2 in **short-circuit transfer** mode can cause the arc to run cold, sluggish, and result in heavy surface scaling and poor toe-line fusion.

- **For Short-Circuit Stainless (Thin Gauge / Sheet Metal):** Switch to a **Helium/Argon/$CO_2$ Tri-Mix** gas (typically 90% He / 7.5% Ar / 2.5% $CO_2$).

- **Why It Matters:** The high helium content drastically increases the thermal energy of the arc, maximizing puddle fluidity, ensuring a flat bead profile, and maintaining the alloy's excellent corrosion resistance without over-oxidizing the weld metal.

**WIRE SELECTION QUICK CHART**

| Job Type | Best Wire | Why |
|---|---|---|
| General steel fab | **ER70S-6** | Best all-around |
| Clean sheet metal | **ER70S-3** | Smooth bead |
| Stainless | **308L / 316L** | Match alloy |
| Stainless to carbon | **309L** | Crack-resistant |
| Aluminum (general) | **4043** | Easy flow |
| Aluminum (marine) | **5356** | Stronger |

**COMMON PROBLEMS & FIXES**

| Problem | Cause | Fix |
|---|---|---|
| Porosity | Dirty metal, wind | Clean, block wind |
| Burn-through | Too hot | Reduce volts/WFS |
| Cold lap | Too cold | Increase heat |
| Bird-nesting | Bad liner, wrong drive rolls | Use U-groove rolls |
| Spatter | Wrong settings | Adjust volts/WFS |

## CHAPTER 21 — FLUX-CORE WIRE (FCAW)

*(Self-Shielded & Gas-Shielded: E71T-11, E71T-1, E71T-8, E81T-1, E91T-1)*

### UNDERSTANDING FLUX-CORE CLASSIFICATIONS

A flux-core classification (like **E71T-11**) tells you:

- **E** = Electrode
- **7** = Tensile strength (70,000 psi)
- **1** = All positions
- **T** = Tubular (flux-core)
- **11** = Self-shielded, general-purpose
- **1** (in E71T-1) = Gas-shielded, all-position

**Self-shielded wires** = outdoor, dirty steel, deep penetration
**Gas-shielded wires** = cleaner, smoother, structural code work

### FLUX-CORE OVERVIEW TABLE

| Wire | Shielding | Strength | Best For |
|---|---|---|---|
| **E71T-11** | Self-shielded | 70 ksi | Outdoor, dirty steel, farm repairs |
| **E71T-1** | Gas-shielded | 70 ksi | Structural, shop fabrication |
| **E71T-8** | Self-shielded | 70 ksi | Heavy structural, deep penetration |
| **E81T-1** | Gas-shielded | 80 ksi | High-strength structural |

| Wire | Shielding | Strength | Best For |
|---|---|---|---|
| **E91T-1** | Gas-shielded | 90 ksi | Heavy equipment, critical welds |

## SELF-SHIELDED WIRES (NO GAS)

These wires are built for **wind, dirt, rust, and field conditions**.

### E71T-11 — THE FARM & FIELD WIRE

**Shielding:** None (self-shielded)
**Strength:** 70,000 psi
**Positions:** All

**Best for**

- Outdoor welding
- Windy conditions
- Dirty, rusty, galvanized steel
- Farm repairs
- Trailers, gates, equipment

**Notes**

- Lots of slag
- More spatter
- Deep penetration
- Great for beginners and field welders

**Settings — 0.035" Wire**

| Thickness | Volts | WFS (ipm) | Notes |
|---|---|---|---|
| 1/8" | 18–19 | 250–300 | Outdoor friendly |
| 3/16" | 19–20 | 300–350 | Deep penetration |
| 1/4" | 20–21 | 350–400 | Multi-pass |

## E71T-8 — THE STRUCTURAL SELF-SHIELDED WIRE

**Shielding:** None
**Strength:** 70,000 psi
**Positions:** All

**Best for**

- Structural steel
- Bridges
- Heavy equipment
- Field erection
- Deep penetration

**Notes**

- Used in code work
- Very deep penetration
- Requires skill
- Slag must be fully removed

**Settings — 0.045" Wire**

| Thickness | Volts | WFS (ipm) |
|---|---|---|
| 1/4" | 20–22 | 180–240 |
| 3/8" | 22–24 | 240–300 |
| 1/2" | 24–26 | 300–360 |

## GAS-SHIELDED WIRES (75/25 OR 90/10)

These wires produce **cleaner, smoother, stronger welds**.

### E71T-1 — THE SHOP STRUCTURAL WIRE

**Shielding:** 75/25 or 90/10
**Strength:** 70,000 psi
**Positions:** All

**Best for**

- Structural fabrication
- Shop work
- Clean steel
- Multi-pass welds

**Notes**

- Smooth bead
- Easy slag removal
- High deposition rate

### CRITICAL OPERATIONAL WARNING: FCAW POLARITY MISMATCH

Before pulling the trigger, always verify your machine's internal polarity. Running flux-cored wire on the wrong polarity is the single most common cause of catastrophic weld porosity, severe spatter, and an erratic, uncontrollable arc in the field.

- Self-Shielded Wire (e.g., E71T-11 / Gasless): Must be run on DCEN (Direct Current Electrode Negative / Straight Polarity).
- Gas-Shielded Wire (e.g., E71T-1 / Dual-Shield): Must be run on DCEP (Direct Current Electrode Positive / Reverse Polarity).

The Field Test: If your puddle is boiling, your shielding gas coverage seems perfect, but your weld looks like a Swiss cheese sponge, stop immediately. Check your ground and gun terminal connections inside the machine door—chances are your polarity is backward for the wire type you are running.

**Settings — 0.045" Wire**

| Thickness | Volts | WFS (ipm) |
|---|---|---|
| 1/4" | 22–24 | 200–260 |
| 3/8" | 24–26 | 260–320 |
| 1/2" | 26–28 | 320–380 |

## E81T-1 — HIGH-STRENGTH STRUCTURAL WIRE

**Shielding:** 75/25 or 90/10
**Strength:** 80,000 psi

**Best for**

- HSLA steels
- Structural beams
- Heavy fabrication
- Bridges

**Settings — 0.045" Wire**

| Thickness | Volts | WFS (ipm) |
|---|---|---|
| 1/4" | 22–24 | 200–260 |
| 3/8" | 24–26 | 260–320 |
| 1/2" | 26–28 | 320–380 |

## E91T-1 — VERY HIGH-STRENGTH WIRE

**Shielding:** 75/25 or 90/10
**Strength:** 90,000 psi

**Best for**

- Heavy equipment
- Crane booms
- High-strength HSLA
- Critical structural welds

**Notes**

- Requires heat control
- Used in demanding applications

**TECHNIQUE NOTES**

**Self-Shielded (E71T-11, E71T-8)**

- **Drag angle** (pull)
- Expect slag — chip thoroughly
- Great for outdoor work
- Keep stickout short

**Gas-Shielded (E71T-1, E81T-1, E91T-1)**

- Push or slight drag
- Cleaner bead
- Less spatter
- Better for shop work

**General FCAW Tips**

- Keep wire stickout **5/8" or less**
- Maintain steady travel speed
- Remove all slag between passes
- Use multi-pass for thick steel

**WIRE SELECTION QUICK CHART**

| Job Type | Best Wire | Why |
|---|---|---|
| Outdoor repairs | **E71T-11** | No gas, wind-proof |
| Structural field work | **E71T-8** | Deep penetration |
| Shop structural | **E71T-1** | Clean, smooth bead |
| High-strength steel | **E81T-1** | 80 ksi strength |
| Heavy equipment | **E91T-1** | Strongest FCAW wire |

**COMMON PROBLEMS & FIXES**

| Problem | Cause | Fix |
|---|---|---|
| Porosity | Wind, dirty metal | Clean, block wind |
| Worm tracks | Too hot | Reduce volts/WFS |
| Slag inclusions | Poor cleaning | Chip between passes |
| Spatter | Wrong angle | Use drag angle |
| Burn-through | Too hot | Reduce heat, faster travel |

## CHAPTER 22 — TIG FILLER RODS (GTAW)

*(Steel, stainless, aluminum, bronze, nickel — what to use, when to use it, and why)*

### UNDERSTANDING TIG FILLER CLASSIFICATIONS

A TIG filler classification (like **ER70S-2**) tells you:

- **ER** = Electrode, Rod
- **70** = Tensile strength (70,000 psi)
- **S** = Solid rod
- **2** = Chemical composition (deoxidizers, alloy content)

For stainless and aluminum, the numbers correspond to alloy families:

- **308L** = 304 stainless
- **316L** = 316 stainless
- **309L** = stainless-to-carbon
- **4043** = aluminum silicon alloy
- **5356** = aluminum magnesium alloy

**TIG FILLER OVERVIEW TABLE**

| Filler | Material | Best For |
|---|---|---|
| **ER70S-2** | Mild steel | Dirty/unknown steel, TIG root passes |
| **ER70S-6** | Mild steel | Clean steel, general fab |
| **308L** | Stainless | 304 stainless |
| **316L** | Stainless | Marine stainless |
| **309L** | Stainless | Stainless to carbon steel |
| **4043** | Aluminum | 3003, 6061, cast aluminum |
| **5356** | Aluminum | Marine aluminum, 5xxx series |
| **ERCuSi-A** | Silicon bronze | Brazing, sheet metal, dissimilar metals |
| **Ni-99 / Ni-55** | Cast iron | Repairs, cracks |
| **ER312** | Tool steels | Unknown alloys, crack-resistant |

## MILD STEEL TIG FILLERS

### ER70S-2 — The "Problem Solver" Steel Rod

**Best for:**

- Dirty steel
- Unknown steel
- TIG root passes
- Pipe welding
- Cast steel repairs

**Notes:**

- Triple-deoxidized
- Handles contamination better
- Smooth puddle

### ER70S-6 — The Clean Steel Rod

**Best for:**

- Clean mild steel
- Sheet metal
- General fabrication

**Notes:**

- Higher deoxidizers than S-2
- Most common steel TIG rod

## STAINLESS TIG FILLERS

### 308L — For 304 Stainless

**Best for:**

- Food-grade stainless
- General stainless fab

**Notes:**

- Most common stainless filler

### 316L — For 316 Stainless

**Best for:**

- Marine stainless
- Chemical environments

**Notes:**

- Better corrosion resistance

### 309L — The Dissimilar Metal Rod

**Best for:**

- Stainless to carbon steel
- Unknown stainless
- Buffer layers

**Notes:**

- The "fix anything" stainless rod

## ALUMINUM TIG FILLERS

### 4043 — The Easy-Flow Aluminum Rod

**Best for:**

- 3003
- 6061
- Cast aluminum

**Notes:**

- Flows beautifully
- Less cracking
- Best all-around aluminum filler

### 5356 — The Strong Aluminum Rod

**Best for:**

- 5052
- 5083
- 5086
- Marine aluminum

**Notes:**

- Stronger than 4043
- Not ideal for anodized color match

## BRONZE & SPECIALTY FILLERS

### ERCuSi-A — Silicon Bronze

**Best for:**

- TIG brazing
- Sheet metal
- Copper, brass, bronze
- Steel to copper
- Decorative work

**Notes:**

- Low heat input
- Beautiful golden bead
- Extremely versatile

**Nickel Rods (Ni-99 / Ni-55)**

**Best for:**

- Cast iron
- Cracked housings
- Engine blocks

**Notes:**

- Ni-99 = machinable
- Ni-55 = stronger

**ER312 — The Tool Steel Rod**

**Best for:**

- Unknown tool steels
- Air-hardening steels
- Crack-sensitive repairs

**Notes:**

- Universal tool-steel filler
- Very crack-resistant

**SETTINGS & PARAMETERS**

These are **real-world TIG ranges** welders actually use.

**STEEL TIG SETTINGS — ER70S-2 / ER70S-6**

| Thickness | Tungsten | Amps | Gas Flow | Notes |
|---|---|---|---|---|
| 1/16" | 1/16" | 40–70 | 10–15 CFH | Clean metal |
| 1/8" | 3/32" | 80–120 | 12–18 CFH | Most common |
| 3/16" | 1/8" | 120–180 | 15–20 CFH | Preheat helps |

## STAINLESS TIG SETTINGS — 308L / 316L / 309L

| Thickness | Tungsten | Amps | Gas Flow | Notes |
|---|---|---|---|---|
| 0.035" | 1/16" | 20–40 | 10–15 CFH | Thin sheet |
| 1/16" | 1/16" | 40–70 | 12–15 CFH | Most common |
| 1/8" | 3/32" | 80–120 | 15–20 CFH | Avoid overheating |

## ALUMINUM TIG SETTINGS — 4043 / 5356 (AC)

| Thickness | Tungsten | Amps | Gas Flow | Notes |
|---|---|---|---|---|
| 1/16" | 1/16" | 50–80 | 12–15 CFH | AC balance ~70% EN |
| 1/8" | 3/32" | 90–140 | 15–20 CFH | Most common |
| 3/16" | 1/8" | 140–200 | 18–22 CFH | Preheat helps |

## SILICON BRONZE TIG SETTINGS — ERCuSi-A

| Thickness | Amps | Notes |
|---|---|---|
| Sheet metal | 40–80 | Low heat, fast travel |
| 1/8" | 80–120 | Great for dissimilar metals |

## TECHNIQUE NOTES

### Steel TIG

- Tight arc
- Add filler consistently
- Clean metal = better welds

### Stainless TIG

- Avoid overheating
- Use back purge for pipe
- Keep heat input low

### Aluminum TIG

- Use AC balance
- Clean oxide layer
- Add filler quickly

### Silicon Bronze

- Low heat
- Excellent for sheet metal
- Great for copper, brass, steel

**FILLER SELECTION QUICK CHART**

| Job Type | Best Filler | Why |
|---|---|---|
| Dirty/unknown steel | **ER70S-2** | Best deoxidizers |
| Clean steel | **ER70S-6** | Smooth bead |
| 304 stainless | **308L** | Alloy match |
| 316 stainless | **316L** | Marine grade |
| Stainless to carbon | **309L** | Crack-resistant |
| Aluminum (general) | **4043** | Easy flow |
| Aluminum (marine) | **5356** | Stronger |
| Cast iron | **Ni-99 / Ni-55** | Crack-resistant |
| Tool steel | **ER312** | Universal tool-steel rod |
| Sheet metal / dissimilar | **Silicon bronze** | Low heat, flexible |

## COMMON PROBLEMS & FIXES

| Problem | Cause | Fix |
|---|---|---|
| Porosity | Dirty metal | Clean thoroughly |
| Sugaring (stainless) | No purge | Add purge gas |
| Collapsing puddle | Too slow | Move faster |
| Cracking | Wrong filler | Use 309L or ER312 |
| Oxide contamination | Dirty aluminum | Clean, adjust AC balance |

## CHAPTER 23 — TUNGSTEN SELECTION & SETUP (GTAW)

*(Pure, Thoriated, Ceriated, Lanthanated, Zirconiated — AC, DC, grinding, cup sizes, gas flow)*

### TUNGSTEN TYPES OVERVIEW

| Tungsten Type | Color | Best For | Notes |
|---|---|---|---|
| **Pure (WP)** | Green | AC aluminum | Old-school, balls easily |
| **2% Thoriated (WT20)** | Red | DC steel/stainless | Most stable DC arc |
| **1.5% Lanthanated (WL15)** | Gold | AC & DC (universal) | Great all-around tungsten |
| **2% Lanthanated (WL20)** | Blue | AC & DC (universal) | Best modern all-purpose |
| **Ceriated (WC20)** | Orange | Low-amp DC | Thin sheet, precision |
| **Zirconiated (WZr)** | Brown | AC aluminum | Resists contamination |

**Recommended all-around tungsten:**
👉 **2% Lanthanated (Blue)** — works for *everything*.

## TUNGSTEN SIZE SELECTION

| Tungsten Size | Amps Range (DC) | Amps Range (AC) | Best Use |
|---|---|---|---|
| **1/16"** | 20–90 | 30–80 | Thin sheet, stainless |
| **3/32"** | 80–180 | 80–160 | Most common size |
| **1/8"** | 150–300 | 150–250 | Heavy aluminum |
| **5/32"** | 250–400 | 250–350 | Thick plate |

**Rule of thumb:**

- **1/16"** = thin
- **3/32"** = everyday fab
- **1/8"** = heavy aluminum

## TUNGSTEN PREPARATION

### DC (Steel, Stainless, Chromoly)

- **Grind to a sharp point**
- Grind **lengthwise**, not sideways
- Slight taper (2–3× tungsten diameter)
- Flat the tip slightly to prevent balling

**Why:**

- Stable arc
- Deep penetration
- Narrow bead

**AC (Aluminum)**

Two options depending on tungsten type:

**Modern method (Lanthanated / Ceriated / Thoriated)**

- **Sharpen like DC**
- Let AC form a **small ball** naturally
- Best arc control and penetration

**Old-school method (Pure / Zirconiated)**

- **Ball the tip intentionally**
- Use AC with high EP to form ball
- Good cleaning action

**Modern practice:**
Use **2% Lanthanated**, sharpened, even for aluminum.

**CUP SIZE SELECTION**

| Cup Size | Gas Flow (CFH) | Best Use |
|---|---|---|
| **#4–#5** | 8–12 | Tight spaces, thin sheet |
| **#6–#7** | 12–18 | General TIG welding |
| **#8** | 15–22 | Stainless, aluminum |

| Cup Size | Gas Flow (CFH) | Best Use |
| --- | --- | --- |
| **#10–#12 (Gas Lens)** | 18–30 | Stainless pipe, titanium |

**Gas lens recommended for:**

- Stainless
- Titanium
- Aluminum
- Anything requiring long stickout

**GAS FLOW SETTINGS**

| Material | Gas | Flow Rate | Notes |
| --- | --- | --- | --- |
| **Steel** | Argon | 12–18 CFH | Standard TIG |
| **Stainless** | Argon | 15–22 CFH | Gas lens helps |
| **Aluminum** | Argon | 15–25 CFH | AC cleaning needs more gas |
| **Titanium** | Argon | 20–35 CFH | Gas lens + trailing shield |

## POLARITY & MACHINE SETTINGS

### DCEN (Direct Current Electrode Negative)

**Used for:**

- Steel
- Stainless
- Chromoly
- Tool steels

**Why:**

- Deep penetration
- Stable arc

### AC (Alternating Current)

**Used for:**

- Aluminum
- Magnesium

**Key settings:**

- **AC Balance:** Start at **70% EN**
    - More EN = more penetration
    - More EP = more cleaning
- **AC Frequency:**
    - 60–120 Hz = general welding
    - 120–200 Hz = tight arc, better control

## TUNGSTEN SELECTION QUICK CHART

| Job Type | Tungsten | Size | Prep |
|---|---|---|---|
| Thin stainless | 2% Lanthanated | 1/16" | Sharp |
| General steel | 2% Lanthanated | 3/32" | Sharp |
| Aluminum (general) | 2% Lanthanated | 3/32" | Sharpened, small ball |
| Aluminum (heavy) | 2% Lanthanated | 1/8" | Sharpened |
| Precision low-amp | Ceriated | 1/16" | Sharp |
| Old-school aluminum | Pure | 1/8" | Balled |

**COMMON PROBLEMS & FIXES**

| Problem | Cause | Fix |
|---|---|---|
| Tungsten balling on DC | Wrong polarity | Switch to DCEN |
| Dirty aluminum welds | Not enough EP | Increase AC balance |
| Arc wandering | Contaminated tungsten | Regrind |
| Porosity | Low gas flow | Increase CFH |
| Black soot on aluminum | Wrong polarity | Use AC |

# SECTION 4 — PROCESS SETUP & SETTINGS CHARTS

## CHAPTER 24 — STICK WELDING SETTINGS (SMAW)

*(Amps by rod size, steel thickness, polarity, and real-world technique notes)*

### HOW TO USE THIS CHAPTER

This chapter gives you:

- **Amperage ranges** for every common rod
- **Thickness recommendations**
- **Polarity requirements**
- **Technique notes**
- **Quick-select charts** for fast field decisions

These are **real-world settings**, not lab numbers — the ranges welders actually use.

### ROD POLARITY QUICK GUIDE

| Rod | Polarity | Notes |
|---|---|---|
| **6010** | DCEP | Deep penetration, fast freeze |
| **6011** | AC / DCEP | Best for dirty steel |
| **6013** | AC / DCEP | Thin sheet, smooth bead |
| **7014** | AC / DCEP | Fast fill, flat/horizontal |
| **7018** | DCEP | Low-hydrogen structural |

| Rod | Polarity | Notes |
|---|---|---|
| **7024** | AC / DCEP | High deposition, flat only |

## AMP SETTINGS BY ROD SIZE

These are the **core settings** welders use in the field.

### 6010 SETTINGS (Deep Penetration, Pipe Roots)

| Rod Size | Amps | Best Use |
|---|---|---|
| 3/32" | 70–90 | Thin steel, open roots |
| 1/8" | 85–120 | General fab, pipe |
| 5/32" | 110–160 | Heavy plate |

**Technique:** Whip/pause, long arc, fast travel.

### 6011 SETTINGS (Farm, Field, Dirty Steel)

| Rod Size | Amps | Best Use |
|---|---|---|
| 3/32" | 60–90 | Thin/galvanized |
| 1/8" | 85–120 | Farm repairs |
| 5/32" | 110–160 | Rusty steel |

**Technique:** Whip/pause, works great on AC buzz boxes.

**6013 SETTINGS (Sheet Metal, Smooth Beads)**

| Rod Size | Amps | Best Use |
|---|---|---|
| 3/32" | 40–70 | Thin sheet |
| 1/8" | 80–110 | Smooth fillets |

**Technique:** Short arc, slow travel, low penetration.

**7014 SETTINGS (Fast Fill, Iron Powder)**

| Rod Size | Amps | Best Use |
|---|---|---|
| 1/8" | 110–150 | Flat fillets |
| 5/32" | 150–200 | Heavy flat welds |

**Technique:** Drag rod, high deposition.

**7018 SETTINGS (Structural, Low-Hydrogen)**

| Rod Size | Amps | Best Use |
|---|---|---|
| 3/32" | 70–100 | Thin structural |
| 1/8" | 115–135 | Most common |
| 5/32" | 150–190 | Heavy plate |

**Technique:** Short arc, drag or slight push, keep rods dry.

**7024 SETTINGS (High-Speed Production)**

| Rod Size | Amps | Best Use |
|---|---|---|
| 1/8" | 140–180 | Flat fillets |
| 5/32" | 180–240 | Heavy flat welds |

**Technique:** Drag only, extremely high deposition.

## AMP SETTINGS BY STEEL THICKNESS

This chart helps welders pick the right rod AND amps based on material thickness.

| Steel Thickness | Recommended Rod | Amps | Notes |
|---|---|---|---|
| 1/16" | 6013 (3/32") | 40–60 | Avoid burn-through |
| 1/8" | 6011 or 7018 (1/8") | 85–120 | Most common thickness |
| 3/16" | 7018 (1/8") | 115–135 | Strong structural |
| 1/4" | 7018 (1/8" or 5/32") | 135–170 | Multi-pass |
| 3/8" | 7018 (5/32") | 150–190 | Preheat helps |
| 1/2" | 7018 (5/32") | 170–200 | Multi-pass, bevel edges |

## QUICK-SELECT ROD CHART

| Job Type | Best Rod | Why |
|---|---|---|
| Dirty/rusty steel | **6011** | Burns through contamination |
| Pipe root | **6010** | Deep penetration |
| Structural steel | **7018** | Low-hydrogen strength |
| Thin sheet | **6013** | Low penetration |
| Flat fillets | **7024** | High deposition |
| General repairs | **6011** | Works on anything |

## TECHNIQUE NOTES

### Arc Length

- Short arc = smoother bead
- Long arc = more penetration (6010/6011 only)

### Travel Speed

- Too slow = undercut, excessive buildup
- Too fast = lack of fusion

### Rod Angle

- Drag for 7018, 7014, 7024
- Whip/pause for 6010/6011
- Straight or slight push for 6013

**Cleaning**

- 7018 requires clean steel
- 6010/6011 tolerate rust, paint, and dirt

**COMMON PROBLEMS & FIXES**

| Problem | Cause | Fix |
|---|---|---|
| Rod sticking | Too cold | Increase amps |
| Porosity | Dirty metal | Clean, increase heat |
| Undercut | Too hot | Reduce amps, slow down |
| Spatter | Long arc | Shorten arc |
| Cracking | Wrong rod | Use 7018 |

## CHAPTER 25 — MIG SETTINGS (GMAW)

*(Volts, wire speed, gas mixes, transfer modes, steel/stainless/aluminum settings)*

### THE MIG SETUP PYRAMID (What Matters Most)

1. **Wire diameter**
2. **Gas mix**
3. **Transfer mode**
4. **Voltage**
5. **Wire feed speed (WFS)**
6. **Stickout & technique**

If the first three are wrong, the last three can't save you.

## GAS MIXES — WHAT THEY DO & WHEN TO USE THEM

| Gas Mix | Best For | Transfer Mode | Notes |
|---|---|---|---|
| **75/25 (C25)** | Mild steel | Short-circuit | Most common MIG gas |
| **100% $CO_2$** | Mild steel | Short-circuit | More penetration, more spatter |
| **90/10** | Mild steel | Spray | Structural, heavy plate |
| **98/2 Ar-$O_2$** | Stainless | Short-circuit & spray | Clean, stable arc |
| **100% Argon** | Aluminum | Spray | Required for aluminum |
| **75/25 He/Ar** | Thick aluminum | Spray | More heat |

**Quick rule:**

- **Steel = 75/25**
- **Stainless = 98/2**
- **Aluminum = 100% argon**

## TRANSFER MODES — THE 4 MODES THAT MATTER

### 1. Short-Circuit (Short Arc)

**Volts:** 16–20
**WFS:** Low–medium
**Best for:**

- Thin steel
- Sheet metal
- Out-of-position welding
- Farm repairs

**Pros:** Low heat, easy to control
**Cons:** Can cold-lap if too cold

**2. Globular**

**Volts:** 20–24
**WFS:** Medium
**Best for:**

- Thick steel
- $CO_2$ gas

**Pros:** Deep penetration
**Cons:** Spatter city — avoid if possible

**3. Spray Transfer**

**Volts:** 24–30
**WFS:** High
**Gas:** 90/10 or 98/2
**Best for:**

- Structural steel
- Heavy plate
- Aluminum

**Pros:** Smooth, hot, deep fusion
**Cons:** Flat/horizontal only

**4. Pulsed Spray**

**Volts:** Machine-controlled
**Best for:**

- Aluminum
- Stainless
- Out-of-position spray

**Pros:** Spray penetration without the heat
**Cons:** Requires advanced machine

**MILD STEEL SETTINGS — ER70S-6**

These are **real-world settings** welders actually use.

**0.023" Wire (Thin Sheet)**

| Thickness | Volts | WFS (ipm) | Mode |
|---|---|---|---|
| 22–20 ga | 14–16 | 180–240 | Short arc |
| 1/16" | 15–16.5 | 200–260 | Short arc |

**0.030" Wire (Most Common)**

| Thickness | Volts | WFS (ipm) | Mode |
|---|---|---|---|
| 16 ga | 16–17 | 200–260 | Short arc |
| 1/8" | 17–18.5 | 260–320 | Short arc |
| 3/16" | 18.5–20 | 320–380 | Hot short arc |
| 1/4" | 20–21 | 380–450 | Short arc / globular |

**0.035" Wire**

| Thickness | Volts | WFS (ipm) | Mode |
|---|---|---|---|
| 1/8" | 17–19 | 200–280 | Short arc |
| 3/16" | 19–20.5 | 280–340 | Hot short arc |
| 1/4" | 20.5–22 | 340–400 | Globular |

**0.045" Wire (Spray Transfer)**

| Thickness | Volts | WFS (ipm) | Mode |
|---|---|---|---|
| 1/4" | 24–26 | 200–260 | Spray |
| 3/8" | 26–28 | 260–320 | Spray |
| 1/2" | 28–30 | 320–380 | Spray |

**STAINLESS SETTINGS — 308L / 316L / 309L**

**0.030" Wire**

| Thickness | Volts | WFS (ipm) | Gas |
|---|---|---|---|
| 16 ga | 16–17 | 200–260 | 98/2 |
| 1/8" | 17–18.5 | 240–300 | 98/2 |
| 3/16" | 18.5–20 | 300–360 | 98/2 |

**ALUMINUM SETTINGS — 4043 / 5356**

**0.030" Wire (Spool Gun)**

| Thickness | Volts | WFS (ipm) | Gas |
|---|---|---|---|
| 1/16" | 17–18 | 350–450 | Argon |
| 1/8" | 18–20 | 450–550 | Argon |
| 3/16" | 20–22 | 550–650 | Argon |

**0.035" Wire**

| Thickness | Volts | WFS (ipm) |
|---|---|---|
| 1/8" | 18–20 | 350–450 |
| 3/16" | 20–22 | 450–550 |
| 1/4" | 22–24 | 550–650 |

## TECHNIQUE NOTES

### Stickout

- Steel: **3/8"**
- Stainless: **1/4–3/8"**
- Aluminum: **1/4" or less**

### Angle

- **Push for everything** except flux-core
- 10–15° push angle

### Travel Speed

- Too slow = undercut, burn-through
- Too fast = cold lap, lack of fusion

### Cleanliness

- Steel: wire brush
- Stainless: stainless brush only
- Aluminum: stainless brush + acetone

## QUICK-SELECT SETTINGS CHART

| Material | Wire | Gas | Volts | WFS |
|---|---|---|---|---|
| Thin steel | 0.023" | 75/25 | 15–16 | 180–240 |
| 1/8" steel | 0.030" | 75/25 | 17–19 | 260–320 |
| 1/4" steel | 0.035" | 75/25 | 20–22 | 340–400 |
| Structural steel | 0.045" | 90/10 | 24–28 | 200–320 |
| Stainless | 0.030" | 98/2 | 17–20 | 240–360 |
| Aluminum | 0.030" | Argon | 18–22 | 450–650 |

## COMMON PROBLEMS & FIXES

| Problem | Cause | Fix |
|---|---|---|
| Porosity | Wind, dirty metal | Clean, block wind |
| Cold lap | Too cold | Increase volts/WFS |
| Burn-through | Too hot | Reduce volts, faster travel |
| Spatter | Wrong settings | Adjust volts/WFS |
| Bird-nesting | Wrong drive rolls | Use U-groove rolls |

## CHAPTER 26 — FLUX-CORE SETTINGS (FCAW)

*(Self-Shielded & Gas-Shielded: E71T-11, E71T-1, E71T-8, E81T-1, E91T-1)*

This chapter is the **settings companion** to Chapter 12 (Flux-Core Wire Types).
These are **real-world field settings**, not lab numbers.

### 1 — SELF-SHIELDED WIRES (NO GAS)

*(Outdoor, dirty steel, wind-proof)*

#### E71T-11 — General Purpose / Farm & Field Wire

**Wire sizes:** 0.030", 0.035", 0.045"
**Polarity:** DCEN
**Shielding:** None
**Best for:** Outdoor, dirty steel, galvanized, farm repairs

**0.030" Wire Settings**

| Thickness | Volts | WFS (ipm) | Notes |
|---|---|---|---|
| 1/16" | 15–16 | 250–300 | Short stitch welds |
| 1/8" | 17–18 | 300–350 | Most common |
| 3/16" | 18–19 | 350–400 | Deep penetration |

**0.035" Wire Settings**

| Thickness | Volts | WFS (ipm) |
| --- | --- | --- |
| 1/8" | 18–19 | 250–300 |
| 3/16" | 19–20 | 300–350 |
| 1/4" | 20–21 | 350–400 |

**E71T-8 — Structural Self-Shielded Wire**

**Wire size:** 0.045"
**Polarity:** DCEN
**Best for:** Structural steel, beams, heavy equipment

**0.045" Wire Settings**

| Thickness | Volts | WFS (ipm) | Notes |
| --- | --- | --- | --- |
| 1/4" | 20–22 | 180–240 | Deep penetration |
| 3/8" | 22–24 | 240–300 | Structural work |
| 1/2" | 24–26 | 300–360 | Multi-pass |

## 2 — GAS-SHIELDED WIRES

*(Cleaner, smoother, structural shop work)*

### E71T-1 — Shop Structural Wire

**Wire size:** 0.045"
**Polarity:** DCEP
**Gas:** 75/25 or 90/10
**Best for:** Structural fabrication, multi-pass welds

**0.045" Wire Settings**

| Thickness | Volts | WFS (ipm) |
|---|---|---|
| 1/4" | 22–24 | 200–260 |
| 3/8" | 24–26 | 260–320 |
| 1/2" | 26–28 | 320–380 |

### E81T-1 — High-Strength Structural Wire

**Wire size:** 0.045"
**Strength:** 80 ksi
**Gas:** 75/25 or 90/10
**Best for:** HSLA steels, beams, bridges

**Settings**

| Thickness | Volts | WFS (ipm) |
|---|---|---|
| 1/4" | 22–24 | 200–260 |
| 3/8" | 24–26 | 260–320 |
| 1/2" | 26–28 | 320–380 |

### E91T-1 — Very High-Strength Wire

**Wire size:** 0.045"
**Strength:** 90 ksi
**Gas:** 75/25 or 90/10
**Best for:** Crane booms, heavy equipment, critical welds

**Settings**

| Thickness | Volts | WFS (ipm) |
|---|---|---|
| 1/4" | 23–25 | 220–280 |
| 3/8" | 25–27 | 280–340 |
| 1/2" | 27–29 | 340–400 |

## TECHNIQUE NOTES (ALL FLUX-CORE)

- **Self-shielded:** drag angle
- **Gas-shielded:** push or slight drag
- Keep stickout **5/8" or less**
- Remove slag between passes
- Avoid weaving with self-shielded wires
- Multi-pass required for anything over 1/4"

## COMMON PROBLEMS & FIXES

| Problem | Cause | Fix |
|---|---|---|
| Worm tracks | Too hot | Reduce volts/WFS |
| Porosity | Wind or dirty metal | Clean, block wind |
| Slag inclusions | Poor cleaning | Chip between passes |
| Burn-through | Too hot | Reduce heat |
| Spatter | Wrong angle | Adjust to drag/push |

## CHAPTER 27 — TIG SETTINGS (GTAW)

*(Steel, stainless, aluminum — amps, tungsten size, gas flow, AC balance, frequency)*

This chapter gives welders **real-world TIG settings** they can trust.
Everything here is based on **practical field ranges**, not lab numbers.

### 1 — STEEL TIG SETTINGS (ER70S-2 / ER70S-6)

**Polarity: DCEN**

**Gas: 100% Argon**

**Cup: #6–#8**

**Tungsten: 2% Lanthanated (recommended)**

| Thickness | Tungsten | Amps | Gas Flow (CFH) | Notes |
|---|---|---|---|---|
| 1/16" | 1/16" | 40–70 | 10–15 | Thin sheet |
| 1/8" | 3/32" | 80–120 | 12–18 | Most common |
| 3/16" | 1/8" | 120–180 | 15–20 | Preheat helps |
| 1/4" | 1/8" | 180–220 | 18–22 | Multi-pass |

**Technique Notes:**

- Tight arc
- Add filler consistently
- Clean metal = better welds
- Use ER70S-2 for dirty/unknown steel

## 2 — STAINLESS TIG SETTINGS (308L / 316L / 309L)

**Polarity: DCEN**

**Gas: 100% Argon**

**Cup: #8–#12 (gas lens recommended)**

**Back Purge: Required for pipe**

| Thickness | Tungsten | Amps | Gas Flow | Notes |
|---|---|---|---|---|
| 0.035" | 1/16" | 20–40 | 10–15 | Thin sheet |
| 1/16" | 1/16" | 40–70 | 12–15 | Most common |
| 1/8" | 3/32" | 80–120 | 15–20 | Avoid overheating |
| 3/16" | 1/8" | 120–160 | 18–22 | Slow travel |

**Technique Notes:**

- Avoid overheating (prevents sugaring)
- Use 309L for stainless-to-carbon
- Use gas lens for long stickout

## 3 — ALUMINUM TIG SETTINGS (4043 / 5356)

**Polarity: AC**

**Gas: 100% Argon**

**Cup: #7–#10**

**Tungsten: 2% Lanthanated (sharpened)**

| Thickness | Tungsten | Amps | Gas Flow | Notes |
| --- | --- | --- | --- | --- |
| 1/16" | 1/16" | 50–80 | 12–15 | AC balance ~70% EN |
| 1/8" | 3/32" | 90–140 | 15–20 | Most common |
| 3/16" | 1/8" | 140–200 | 18–22 | Preheat helps |
| 1/4" | 1/8" | 200–260 | 20–25 | Heavy plate |

**AC Settings:**

- **AC Balance:** 70% EN (more EN = more penetration)
- **AC Frequency:**
  - 60–120 Hz = general welding
  - 120–200 Hz = tight arc, better control

**Technique Notes:**

- Clean oxide layer
- Add filler quickly
- Keep arc tight
- Use 4043 for easy flow, 5356 for strength

## 4 — SPECIALTY SETTINGS

### Silicon Bronze (ERCuSi-A)

- Amps: 40–120
- Low heat
- Great for sheet metal and dissimilar metals

### Cast Iron (Ni-99 / Ni-55)

- Amps: 70–150
- Preheat recommended
- Slow cool

# CHAPTER 28 — MACHINE SETUP & TROUBLESHOOTING

*(Voltage, inductance, arc force, hot start, pulse, slope, common machine issues)*

This chapter teaches welders how to **dial in the machine itself**, not just the filler and technique.

## 1 — MIG MACHINE CONTROLS

### Voltage

- Controls arc length
- Too low = cold lap
- Too high = spatter, undercut

### Wire Feed Speed (WFS)

- Controls amperage
- Too low = stubbing
- Too high = burn-back

### Inductance

- Higher = smoother arc, less spatter
- Lower = crisp arc, more penetration

### Pre-Flow / Post-Flow

- Pre-flow: 0.2–0.5 sec
- Post-flow: 2–5 sec

## 2 — STICK MACHINE CONTROLS

### Arc Force (Dig)

- Higher = more penetration, prevents sticking
- Lower = softer arc

### Hot Start

- Boosts amps at arc strike
- Helps prevent sticking

### Polarity

- DCEP for 7018, 6010
- AC for 6011, 6013

## 3 — TIG MACHINE CONTROLS

### AC Balance

- More EN = deeper penetration
- More EP = more cleaning

### AC Frequency

- Low (60–90 Hz): wide arc
- High (120–200 Hz): tight arc

### Slope Up / Slope Down

- Slope up: prevents arc blowout
- Slope down: prevents crater cracking

**Pulse**

- Helps with thin metal
- Reduces heat input
- Improves bead appearance

## 4 — COMMON MACHINE PROBLEMS & FIXES

| **Problem** | **Cause** | **Fix** |
|---|---|---|
| Porosity | Gas issue | Check flow, leaks, wind |
| Burn-back | WFS too high | Reduce WFS |
| Sticking rod | Amps too low | Increase amps |
| Arc wandering | Contaminated tungsten | Regrind |
| Spatter | Wrong volts/WFS | Adjust settings |
| Cold lap | Too cold | Increase volts/WFS |
| Overheating | Duty cycle | Let machine cool |

## 5 — GAS TROUBLESHOOTIN

### Too Much Gas

- Turbulence
- Sucks in air
- Causes porosity

### Too Little Gas

- Oxidation
- Black soot (aluminum)
- Sugaring (stainless)

### Leaks

- Check hoses
- Check regulator
- Check torch fittings

## CHAPTER 29 — JOINT TYPES & PREPARATION

*(Butt, lap, tee, corner, edge, bevels, gaps, fit-up, cleaning)*

Joint prep is where **90% of weld quality is decided.**
This chapter gives welders the exact bevels, gaps, and prep steps for every joint.

### THE 5 FUNDAMENTAL JOINT TYPES

| Joint Type | Where used | Notes |
|---|---|---|
| **Butt Joint** | Plate, pipe | Requires bevels on thicker material |
| **Lap Joint** | Sheet metal | Strong, easy to weld |
| **Tee Joint** | Frames, structural | Fillet welds |
| **Corner Joint** | Boxes, frames | Inside/outside corners |
| **Edge Joint** | Sheet metal | Low strength, cosmetic |

### BUTT JOINTS

#### Under 1/8" Thick

- No bevel
- Tight fit-up
- MIG/TIG: 1/32" gap helps penetration
- Stick: no gap needed

**1/8"–3/8" Thick**

- **Single bevel**
- 30–37.5° bevel angle
- 1/16" root face
- 1/16"–1/8" root opening

**Over 3/8" Thick**

- **Double bevel** or **double V**
- Reduces distortion
- Required for structural work

**LAP JOINTS**

**Rules**

- Overlap at least **1× material thickness**
- Fillet weld on one or both sides
- MIG/TIG: push
- Stick: drag

**Uses**

- Sheet metal
- Brackets
- Frames

## TEE JOINTS

### Fillet Welds

- Most common weld in fabrication
- Leg size ≈ material thickness

### Fit-Up

- Tight joint
- Tack both sides
- Back-step to reduce distortion

## CORNER JOINTS

### Outside Corner

- Used for boxes, frames
- MIG/TIG: push
- Stick: short arc

### Inside Corner

- Stronger
- Easier to weld
- Fillet weld

## EDGE JOINTS

### Where Used

- Sheet metal
- HVAC
- Tanks

### Notes

- Low strength
- TIG preferred
- MIG can blow through thin edges

## BEVEL ANGLES & ROOT OPENINGS

| Thickness | Bevel Angle | Root Face | Root Opening |
|---|---|---|---|
| 1/8" | 30° | 1/32" | 1/16" |
| 3/16" | 30–37.5° | 1/16" | 1/8" |
| 1/4" | 37.5° | 1/16" | 1/8" |
| 3/8"+ | 37.5° | 1/8" | 1/8"–3/16" |

## CLEANING & SURFACE PREP

### Steel

- Grind to shiny metal
- Remove rust, paint, mill scale

### Stainless

- Stainless brush only
- Acetone wipe
- Avoid contamination

### Aluminum

- Stainless brush
- Remove oxide
- Acetone wipe

### Galvanized

- Grind zinc off 1–2"
- Ventilate heavily

## TACK WELDING

### Rules

- Tack every **2–4 inches**
- Opposing tacks prevent pulling
- Pipe: tack at **12, 3, 6, 9 o'clock**

### Size

- 1/4"–3/8"
- Same heat as main weld

## CHAPTER 30 — WELDING POSITIONS (1G–6G)

*(Flat, horizontal, vertical, overhead, pipe positions)*

This chapter teaches welders how to weld in every position used in structural, fabrication, and pipe work.

### THE 4 PLATE POSITIONS

| Position | Name | Description |
|---|---|---|
| **1G** | Flat | Easiest, highest deposition |
| **2G** | Horizontal | Side-to-side control |
| **3G** | Vertical | Up or down |
| **4G** | Overhead | Most difficult plate position |

### THE 3 PIPE POSITIONS

| Position | Name | Description |
|---|---|---|
| **1G** | Rolled | Pipe rotates under torch |
| **2G** | Fixed horizontal | Pipe fixed, weld around |
| **5G** | Fixed vertical | Pipe fixed, weld uphill |
| **6G** | 45° fixed | **Certification position** |

6G is the **hardest** and qualifies welders for almost everything.

## POSITION DETAILS

### 1G — FLAT POSITION

#### Characteristics

- Easiest
- Highest deposition
- Gravity helps puddle

#### Technique

- MIG: push
- Stick: drag
- TIG: tight arc

### 2G — HORIZONTAL POSITION

#### Challenges

- Puddle wants to sag
- Undercut risk

#### Technique

- Slight upward rod/gun angle
- Control travel speed
- Use stringer beads

## 3G — VERTICAL POSITION

### Two Methods

- **Vertical Up (stronger)**
- **Vertical Down (thin metal)**

### Vertical Up Technique

- Small weave or triangle pattern
- Pause at toes
- Best for structural

### Vertical Down Technique

- Fast travel
- Low heat
- Best for sheet metal

## 4G — OVERHEAD POSITION

### Challenges

- Gravity works against you
- Spatter and drip risk

### Technique

- Short arc
- Fast travel
- Small stringers
- Keep puddle small

## PIPE POSITIONS

### 1G PIPE — ROLLED

- Pipe rotates
- Welder stays stationary
- Easiest pipe position

### 2G PIPE — FIXED HORIZONTAL

- Pipe fixed
- Weld around circumference
- Requires side-to-side control

### 5G PIPE — FIXED VERTICAL

- Pipe fixed
- Weld uphill
- Requires vertical-up technique

### 6G PIPE — 45° FIXED (CERTIFICATION)

**Why it's the hardest**

- Combines flat, horizontal, vertical, overhead
- Requires full control of puddle
- Used for pipe certification tests

**Technique**

- Root: keyhole control
- Hot pass: fill undercut
- Fill/cap: weave or stringers

# CHAPTER 31 — DISTORTION CONTROL & HEAT MANAGEMENT

*(How to keep metal straight, flat, and crack-free)*

Distortion is caused by **uneven heating and cooling.**
This chapter gives welders the **exact techniques** to control heat, prevent warping, and keep assemblies' square.

## 1 — WHY DISTORTION HAPPENS

When metal heats up, it **expands**.
When it cools, it **shrinks**.

If one side shrinks more than the other, the part **pulls, twists**, or **cups**.

**Major causes:**

- Too much heat input
- Long welds without breaks
- No tacks or poor fit-up
- Welding only one side
- Thin material
- High-heat processes (stick, flux-core, spray MIG)

## 2 — DISTORTION CONTROL TECHNIQUES

These are the **core techniques** used in fabrication shops and structural work.

### 1. Back-Step Welding

Instead of welding left-to-right:

- Start a short weld
- Stop
- Move backward
- Start the next weld overlapping the previous stop

**Effect:**
Shrinks the metal in the opposite direction → reduces pull.

### 2. Skip Welding

Weld in **short sections**, spaced apart:

- Weld 2–3 inches
- Skip ahead
- Weld another 2–3 inches
- Come back later and fill the gaps

**Effect:**
Spreads heat out → prevents long heat buildup.

### 3. Opposing Welds

Weld one side, then weld the opposite side.

**Effect:**
Equalizes shrinkage → keeps parts straight.

**4. Strong Tack Welds**

Tacks should be:

- 1/4"–3/8" long
- Same heat as main weld
- Placed every 2–4 inches

**Effect:**
Locks the joint in place → prevents movement.

**5. Clamping & Fixturing**

Use:

- C-clamps
- Bar clamps
- Magnetic squares
- Angle fixtures
- Welding tables with dog holes

**Effect:**
Physically restrains the metal.

**6. Heat Sinks**

Use:

- Copper bars
- Aluminum blocks
- Thick steel plates

**Effect:**
Pulls heat away → reduces warping.

**7. Preheat (for thick or hard steels)**

Preheat helps:

- Reduce cracking
- Slow cooling
- Equalize heat distribution

**Typical preheat:**

- Mild steel: 150–250°F
- Tool steel: 400–1000°F
- Cast iron: 400–600°F

**8. Shorter Welds**

Short beads = less heat = less distortion.

## 3 — DISTORTION IN SPECIFIC MATERIALS

**Steel**

- Moderate distortion
- Controlled with tacks and skip welding

**Stainless**

- High distortion
- Use low heat, back-step, and clamps

**Aluminum**

- Very high distortion
- Use heat sinks, fast travel, and skip welding

## 4 — COMMON PROBLEMS & FIXES

| Problem | Cause | Fix |
|---|---|---|
| Warping | Too much heat | Skip weld, back-step |
| Cupping | One-sided weld | Opposing welds |
| Twist | Uneven tacks | Add more tacks |
| Cracking | Cooling too fast | Preheat, slow cool |
| Pulling | Long welds | Short beads, clamps |

## CHAPTER 32 — WELDING SAFETY (Fumes, PPE, Ventilation)

*(Galvanized fumes, stainless hex-chrome, aluminum oxide, UV exposure, PPE)*

This chapter covers the **real hazards** welders face and how to avoid them.

### 1 — FUME HAZARDS

Different metals produce different fumes.

#### Galvanized Steel (Zinc Oxide)

**Hazard:** Metal fume fever
**Symptoms:** Flu-like, chills, nausea
**Prevention:**

- Grind zinc off 1–2"
- Ventilation
- Respirator if needed

#### Stainless Steel (Hexavalent Chromium)

**Hazard:** Hex-chrome (carcinogenic)
**Prevention:**

- Ventilation
- Fume extraction
- Respirator for enclosed spaces

**Aluminum (Aluminum Oxide)**

**Hazard:** Respiratory irritation
**Prevention:**

- Clean metal
- Good airflow

**Manganese (From Steel Wires)**

**Hazard:** Neurological effects
**Prevention:**

- Ventilation
- Avoid welding in confined spaces

## 2 — PPE (Personal Protective Equipment)

**Helmet**

- Auto-darkening
- Shade 10–13 for arc welding
- Shade 8–10 for TIG

**Gloves**

- MIG gloves: thicker
- TIG gloves: thin goatskin
- Stick gloves: heavy leather

**Clothing**

- 100% cotton or leather
- No synthetics
- Welding jacket or sleeves

**Boots**

- Leather
- Steel toe optional
- No mesh or fabric

## 3 — VENTILATION

**Best Practices**

- Use fume extractors
- Keep head out of the plume
- Use fans to move air *across*, not *toward* the weld
- For stainless: use local exhaust

**Confined Spaces**

- Never weld without ventilation
- Use respirator
- Monitor oxygen levels

## 4 — UV & IR EXPOSURE

### Hazards

- Arc flash
- Skin burns
- Eye damage

### Prevention

- Proper helmet
- Welding curtains
- Long sleeves
- Gloves

## 5 — FIRE SAFETY

### Checklist

- Remove flammables
- Keep extinguisher nearby
- Watch for sparks traveling 20–30 feet
- Never weld near fuel tanks without purging

## 6 — ELECTRICAL SAFETY

- Keep gloves dry
- Inspect cables
- Avoid standing in water
- Use proper grounding

## CHAPTER 33 — TROUBLESHOOTING GUIDE

*(Porosity, undercut, lack of fusion, cracking, arc blow, spatter, burn-through)*

This chapter is the **diagnostic manual** welders flip to when something goes wrong.
Every defect is paired with **causes** and **fast, practical fixes**.

### 1 — POROSITY

**What it looks like:**

Tiny holes in the weld bead or inside the weld.

**Causes**

- Dirty metal
- Wind blowing away shielding gas
- Low gas flow
- Gas leaks
- Contaminated filler
- Too long stickout (MIG)
- Moisture in flux (stick/flux-core)

**Fixes**

- Clean metal to shiny
- Increase gas flow 2–5 CFH
- Block wind
- Shorten stickout

- Check hoses and fittings
- Bake 7018 rods if needed

**2 — UNDERCUT**

**What it looks like:**

Groove melted along the toe of the weld.

**Causes**

- Too hot
- Too fast travel
- Wrong angle
- Long arc length

**Fixes**

- Reduce amps/volts
- Slow down
- Adjust angle to 10–15° push/drag
- Shorten arc

**3 — LACK OF FUSION**

**What it looks like:**

Weld sits on top of metal without bonding.

**Causes**

- Too cold

- Travel too fast
- Wrong angle
- Dirty metal
- Incorrect bevel or gap

**Fixes**

- Increase heat
- Slow travel
- Clean metal
- Increase bevel angle
- Add root opening

**4 — CRACKING**

**Types**

- Hot cracks
- Cold cracks
- Crater cracks

**Causes**

- High carbon steel
- No preheat
- Rapid cooling
- Wrong filler
- Ending arc too quickly

**Fixes**

- Preheat (150–600°F depending on metal)
- Slow cool
- Use correct filler (309L, ER312, Ni-99)
- Use crater fill or slope-down

**5 — SPATTER**

**Causes**

- Wrong MIG settings
- Too long stickout
- Dirty metal
- $CO_2$ gas only

**Fixes**

- Adjust volts/WFS
- Shorten stickout
- Clean metal
- Switch to 75/25

## 6 — BURN-THROUGH

**Causes**

- Too hot
- Thin metal
- Too slow
- Too big a gap

**Fixes**

- Reduce heat
- Increase travel speed
- Use stitch welding
- Reduce gap

## 7 — ARC BLOW (Stick & Flux-Core)

**Causes**

- Magnetic fields in steel
- Wrong ground location
- DC welding on thick plate

**Fixes**

- Move ground clamp
- Weld toward the ground
- Switch to AC if possible

## 8 — WANDERING ARC (TIG)

### Causes

- Contaminated tungsten
- Wrong tungsten prep
- Gas turbulence
- Too long stickout

### Fixes

- Regrind tungsten
- Use gas lens
- Reduce gas flow
- Shorten stickout

# CHAPTER 34 — FIELD REPAIRS & EMERGENCY TECHNIQUES

*(Farm equipment, cracked housings, broken bolts, exhausts, cast iron, thin metal)*

This chapter teaches welders how to fix things **fast**, **strong**, and **in the real world** — not in a perfect shop.

## 1 — FARM & EQUIPMENT REPAIRS

**Best rods/wires**

- **6011** — dirty steel, rust, paint
- **7018** — strength
- **E71T-11** — outdoor, wind-proof
- **ER70S-6** — MIG general repairs

**Common repairs**

- Gates
- Trailers
- Loader buckets
- Brackets
- Hinges

**Tips**

- Grind to shiny metal
- Use 6011 for root, 7018 for cap
- Preheat thick steel

## 2 — CRACKED HOUSINGS (CAST IRON)

**Best fillers**

- **Ni-99** (machinable)
- **Ni-55** (stronger)
- **Silicon bronze** (low heat)

**Repair method**

1. Drill stop-holes at crack ends
2. V-groove the crack
3. Preheat 300–600°F
4. Weld short beads
5. Peen lightly
6. Slow cool in sand or blanket

## 3 — BROKEN BOLTS

**Method 1 — Weld a Nut On**

1. Clean bolt face
2. Place nut over bolt
3. Fill nut with weld
4. Let cool slightly
5. Turn out with wrench

**Method 2 — Build Up & Grab**

- Build weld metal on broken stud
- Grab with vise-grips

## 4 — EXHAUST & THIN METAL

**Best processes**

- MIG with 0.023" wire
- TIG with low amps
- 6013 stick (if needed)

**Techniques**

- Stitch weld
- Skip weld
- Backing bars
- Low heat

## 5 — BROKEN FRAMES & STRUCTURAL REPAIRS

**Best rods/wires**

- **7018**
- **E71T-1**
- **ER70S-6**

**Steps**

1. Grind out crack
2. Drill stop-holes
3. Bevel edges
4. Weld root
5. Add reinforcement plate if needed

## 6 — EMERGENCY OUTDOOR REPAIRS

**When wind is bad**

- Use **6011** or **E71T-11**
- Shield weld with body or board

**When metal is dirty**

- Use 6011
- Burn through rust/paint
- Clean after root

**When metal is wet**

- Dry area
- Increase heat
- Use 6011

## 7 — TEMPORARY FIXES (GET-YOU-HOME REPAIRS)

### Examples

- Weld chain links
- Patch brackets
- Quick gussets
- Temporary braces

### Rules

- Mark temporary repairs
- Replace or redo properly later

## CHAPTER 35 — METAL IDENTIFICATION CHARTS

*(Spark tests, magnet tests, hardness clues, color, weight, behavior under grinder)*

This chapter teaches welders how to identify metals **without lab equipment** — using only sight, sound, sparks, weight, magnetism, and behavior under heat.

### 1 — SPARK TEST CHART

Spark testing is one of the fastest ways to identify unknown steels.

| Metal | Spark Color | Spark Pattern | Notes |
|---|---|---|---|
| **Mild Steel** | Bright yellow | Long, straight sparks | Most common |
| **High Carbon Steel** | Bright white | Lots of branching | Knives, springs |
| **Cast Iron** | Dull red/orange | Short, bursty sparks | Brittle |
| **Stainless Steel** | Very few sparks | Dull red | High chromium |
| **Tool Steel** | Bright white | Dense, explosive | Hardened tools |
| **Titanium** | Bright white | Fireworks-like | Very reactive |
| **Nickel Alloys** | Very few sparks | Dull | Inconel, Monel |

## 2 — MAGNET TEST

| Metal | Magnetic? | Notes |
|---|---|---|
| **Mild Steel** | Yes | Strong attraction |
| **High Carbon Steel** | Yes | Strong attraction |
| **Stainless 304** | No | Most common stainless |
| **Stainless 316** | No | Marine grade |
| **Stainless 409/430** | Slight | Automotive exhaust |
| **Cast Iron** | Yes | Heavy, brittle |
| **Aluminum** | No | Lightweight |
| **Copper/Brass/Bronze** | No | Non-ferrous |
| **Titanium** | No | Very light, strong |

## 3 — COLOR & SURFACE IDENTIFICATION

| Metal | Color | Surface Clues |
|---|---|---|
| **Mild Steel** | Dark gray | Rusts easily |
| **Stainless Steel** | Silver/gray | Doesn't rust, smooth |
| **Aluminum** | Bright silver | Very light |
| **Cast Iron** | Dark gray/black | Grainy texture |
| **Copper** | Reddish | Tarnishes brown/green |
| **Brass** | Yellow/gold | Smooth, heavy |
| **Bronze** | Brown/gold | Harder than brass |
| **Titanium** | Dull silver | Very light, strong |

## 4 — HARDNESS & FILE TEST

| Metal | File Behavior | Notes |
|---|---|---|
| **Mild Steel** | Cuts easily | Soft |
| **High Carbon Steel** | Skates or bites shallow | Hard |
| **Tool Steel** | Skates | Very hard |
| **Cast Iron** | Crumbles | Brittle |
| **Aluminum** | Cuts very easily | Soft |
| **Stainless** | Medium resistance | Tough |

## 5 — WEIGHT TEST

| Metal | Relative Weight | Notes |
|---|---|---|
| Aluminum | Very light | 1/3 weight of steel |
| Steel | Heavy | Standard baseline |
| Stainless | Slightly heavier than steel | High density |
| Copper | Very heavy | Dense |
| Brass/Bronze | Heavy | Non-ferrous alloys |
| Titanium | Light | Stronger than steel |

## 6 — HEAT REACTION TEST

| Metal | Behavior Under Heat | Notes |
|---|---|---|
| Aluminum | Melts suddenly | No color change |
| Steel | Glows red → orange → yellow | Predictable |
| Stainless | Glows slowly | Low conductivity |
| Cast Iron | Cracks if overheated | Brittle |
| Titanium | Sparks violently in air | Reactive |

## CHAPTER 36 — PREHEAT & INTERPASS MASTER TABLE

*(All metals in one place — steel, stainless, aluminum, cast iron, tool steels)*

This is one of the most important reference charts in the entire book.
It tells welders **exactly how hot to preheat** and **how hot to keep the metal between passes**.

### 1 — MILD STEEL PREHEAT

| Thickness | Preheat | Interpass Max | Notes |
|---|---|---|---|
| Under 1/4" | None | 300°F | Clean metal |
| 1/4"–1/2" | 150–250°F | 400°F | Reduces cracking |
| Over 1/2" | 250–350°F | 450°F | Heavy plate |

### 2 — STAINLESS STEEL PREHEAT

| All Thicknesses | Preheat | Interpass Max | Notes |
|---|---|---|---|
| 304 / 316 | None | 350°F | Avoid carbide precipitation |
| 409 / 430 | 150°F | 300°F | Automotive stainless |

## 3 — ALUMINUM PREHEAT

| Thickness | Preheat | Interpass Max | Notes |
|---|---|---|---|
| Under 1/8" | None | 250°F | Avoid distortion |
| 1/8"–1/4" | 150–250°F | 300°F | Improves flow |
| Over 1/4" | 250–350°F | 350°F | Required for heavy plate |

## 4 — CAST IRON PREHEAT

| Type | Preheat | Interpass Max | Notes |
|---|---|---|---|
| Gray Iron | 400–600°F | 600°F | Slow cool required |
| Ductile Iron | 300–500°F | 500°F | Ni-55 or Ni-99 |
| Unknown Cast | 400–600°F | 600°F | Safe range |

## 5 — TOOL STEEL PREHEAT

| Tool Steel Type | Preheat | Interpass Max | Notes |
| --- | --- | --- | --- |
| W-Series | **Do not weld** | — | Replace or braze |
| O-Series | 400–600°F | 600°F | Slow cool |
| A-Series | 800–1000°F | 1000°F | Very crack-sensitive |
| D-Series | 800–1000°F | 1000°F | Often brazed |
| H-Series | 600–900°F | 900°F | Hot-work steels |
| S-Series | 400–600°F | 600°F | Impact steels |
| High-Speed Steel | **Do not weld** | — | Replace or braze |

## 6 — HIGH-STRENGTH STEELS (HSLA)

| Grade | Preheat | Interpass Max | Notes |
| --- | --- | --- | --- |
| A36 | 150°F | 400°F | Standard structural |
| A572 | 150–250°F | 450°F | HSLA |
| A514 (T-1) | 200–400°F | 450°F | Heat-sensitive |
| AR Plate | 200–400°F | 450°F | Abrasion-resistant |

## 7 — DISSIMILAR METALS

| Combination | Preheat | Notes |
| --- | --- | --- |
| Stainless → Carbon | 150–250°F | Use 309L |
| Tool Steel → Mild Steel | 400–800°F | Use ER312 |
| Cast Iron → Steel | 300–600°F | Use Ni-99 or Ni-55 |

## CHAPTER 37 — FILLER METAL CROSS-REFERENCE TABLE

*(Stick → MIG → TIG → Flux-Core equivalents)*

This chapter gives welders a **unified chart** showing which filler metals match across all processes.
Perfect for switching machines, matching procedures, or replacing unavailable rods/wire.

### 1 — MILD STEEL (70,000 PSI CLASS)

| Process | Filler Metal | Notes |
|---|---|---|
| **Stick** | 7018 | Structural, low-hydrogen |
| **Stick** | 7014 | Fast-fill, flat/horizontal |
| **Stick** | 6010 / 6011 | Deep penetration, dirty steel |
| **MIG** | ER70S-6 | Most common MIG wire |
| **MIG** | ER70S-3 | Clean steel, sheet metal |
| **TIG** | ER70S-2 | Dirty/unknown steel, root passes |
| **TIG** | ER70S-6 | Clean steel |
| **Flux-Core (Gas)** | E71T-1 | Structural, smooth bead |
| **Flux-Core (Self-Shielded)** | E71T-11 | Outdoor, dirty steel |

**Cross-Reference Summary:**

- **7018 ↔ ER70S-6 ↔ ER70S-2 ↔ E71T-1**
- **6010/6011 ↔ ER70S-2 ↔ E71T-11**

## 2 — STAINLESS STEEL

| Process | Filler Metal | Best For |
|---|---|---|
| **Stick** | 308L-16 | 304 stainless |
| **Stick** | 316L-16 | 316 stainless |
| **Stick** | 309L-16 | Stainless → carbon |
| **MIG** | ER308L | 304 stainless |
| **MIG** | ER316L | Marine stainless |
| **MIG** | ER309L | Dissimilar metals |
| **TIG** | ER308L | 304 stainless |
| **TIG** | ER316L | Marine stainless |
| **TIG** | ER309L | Stainless → carbon |

**Cross-Reference Summary:**

- **308L-16 ↔ ER308L (MIG/TIG)**
- **316L-16 ↔ ER316L (MIG/TIG)**
- **309L-16 ↔ ER309L (MIG/TIG)**

## 3 — ALUMINUM

| Process | Filler Metal | Best For |
|---|---|---|
| MIG | 4043 | 3003, 6061, cast |
| MIG | 5356 | Marine aluminum |
| TIG | 4043 | Easy flow |
| TIG | 5356 | Stronger, stiffer |

**Cross-Reference Summary:**

- **4043 MIG ↔ 4043 TIG**
- **5356 MIG ↔ 5356 TIG**

## 4 — CAST IRON

| Process | Filler Metal | Notes |
|---|---|---|
| Stick | Ni-99 | Machinable |
| Stick | Ni-55 | Stronger |
| TIG | Ni-99 | Cleanest method |
| TIG | Silicon Bronze | Low-heat repairs |

## 5 — TOOL STEELS

| Process | Filler Metal | Notes |
|---|---|---|
| **TIG** | ER312 | Universal tool-steel rod |
| **TIG** | ER309L | For unknown alloys |
| **Stick** | 312-16 | Rare but available |

## 6 — DISSIMILAR METALS

| Combination | Best Filler | Process |
|---|---|---|
| Stainless → Carbon | 309L | MIG/TIG/Stick |
| Tool Steel → Mild Steel | ER312 | TIG |
| Cast Iron → Steel | Ni-99 / Ni-55 | Stick/TIG |
| Steel → Copper/Brass | Silicon Bronze | TIG |

## CHAPTER 38 — WELDING SYMBOLS & BLUEPRINT BASICS

*(Fillet, groove, bevel, root opening, tail notes, arrow side vs other side)*

This chapter teaches welders how to **read and interpret welding symbols**, the language of fabrication and structural work.

### 1 — THE WELDING SYMBOL STRUCTURE

A welding symbol has **three main parts**:

1. **Reference line** — the baseline
2. **Arrow** — points to the joint
3. **Tail** — optional notes (process, filler, etc.)

### 2 — ARROW SIDE VS OTHER SIDE

| Placement | Meaning |
|---|---|
| **Symbol BELOW the line** | Weld on **arrow side** |
| **Symbol ABOVE the line** | Weld on **other side** |
| **Both sides** | Weld both sides |

## 3 — COMMON WELD SYMBOLS

| Symbol | Name | Meaning |
|---|---|---|
| ∟ | **Fillet Weld** | Most common weld |
| V | **V-Groove** | Single bevel |
| X | **Double-V Groove** | Both sides beveled |
| U | **U-Groove** | Thick plate |
| J | **J-Groove** | One-sided U |
| ⌒ | **Plug/Slot Weld** | Holes filled |
| ▭ | **Backing Bar** | Backing required |
| —○— | **Spot Weld** | Resistance weld |
| —●— | **Projection Weld** | Studs, nuts |

## 4 — FILLET WELD SYMBOL DETAILS

### Size

- Shown to the **left** of the symbol
- Example: ¼ ⌊ = 1/4" fillet

### Length

- Shown to the **right** of the symbol
- Example: ⌊ **3-0** = 3" long

### Intermittent Fillets

- Example: ¼ ⌊ **2-4**
    - 1/4" fillet
    - 2" weld
    - 4" pitch

## 5 — GROOVE WELD SYMBOL DETAILS

### Root Opening

- Shown inside the groove symbol
- Example: **V 1/8** = 1/8" root gap

### Bevel Angle

- Shown in tail or near symbol
- Example: **37.5°**

### Backing

- Rectangle symbol under groove
- Means backing bar required

## 6 — TAIL NOTES

The tail can specify:

- Process (SMAW, GMAW, GTAW, FCAW)
- Filler metal (7018, ER70S-6, etc.)
- Preheat requirements
- Weld procedure number (WPS-###)
- Special instructions

Example:
**GMAW-P, ER70S-6, 75/25 gas**

## 7 — FINISHING SYMBOLS

| Symbol | Meaning |
|---|---|
| **C** | Chip |
| **G** | Grind |
| **M** | Machine |
| **R** | Roll |

Example:
**⌞ G** = Fillet weld must be ground smooth.

## 8 — WELD ALL AROUND SYMBOL

A small **circle** at the junction of the arrow and reference line means:

**Weld completely around the joint**

Used for:

- Pipes
- Posts
- Tubes
- Round parts

## 9 — FIELD WELD SYMBOL

A small **flag** on the reference line means:

**Weld must be done in the field (not in shop)**

## CHAPTER 39 — Tools, Consumables & Shop Setup

*(Essential tools, grinders, clamps, PPE, gas cylinders, storage, layout, workflow)*

This chapter gives welders the **complete setup guide** for a functional, efficient, and safe welding shop — whether it's a garage, farm shop, or full fabrication space.

### 1 — ESSENTIAL WELDING TOOLS

These are the tools every welder should have, regardless of process.

**Core Tools**

- Angle grinder (4.5" or 5")
- Welding helmet (auto-darkening)
- MIG pliers
- Chipping hammer
- Wire brushes (steel & stainless)
- Soapstone or paint marker
- Tape measure
- Square (speed square + framing square)
- Clamps (C-clamps, locking pliers, bar clamps)
- Welding magnets
- Scribe & center punch

**Consumables**

- Grinding discs (36–60 grit)
- Flap discs (60–120 grit)
- Cut-off wheels
- Wire wheels
- Nozzle gel (MIG)
- Anti-spatter spray
- Contact tips (MIG)
- Tungsten electrodes (TIG)
- Stick electrodes (various types)

## 2 — GRINDERS & ABRASIVES

### Angle Grinder Types

- **4.5" grinder** — most common
- **5" grinder** — more power
- **Variable speed** — ideal for TIG prep

### Abrasive Types

| Abrasive | Use |
|---|---|
| **Cut-off wheel** | Cutting steel, bolts, plate |
| **Grinding wheel** | Heavy material removal |
| **Flap disc** | Smoothing, blending |
| **Wire wheel** | Rust/paint removal |
| **Sanding disc** | Fine finishing |

### Pro Tips

- Use **flap discs** for weld cleanup
- Use **wire wheels** for rust/paint
- Keep a dedicated **stainless brush** for stainless

## 3 — CLAMPS & FIXTURING

### Clamps You Need

- 6" and 11" locking pliers
- 4"–12" C-clamps
- Bar clamps
- Pipe clamps
- Magnetic squares
- Corner clamps

### Fixturing Tools

- Welding table with dog holes
- Angle plates
- Stops and squares
- Grounding tabs

### Why Fixturing Matters

- Keeps parts square
- Reduces distortion
- Speeds up production

## 4 — PPE (PERSONAL PROTECTIVE EQUIPMENT)

### Helmet

- Auto-darkening
- Shade 10–13 for MIG/Stick
- Shade 8–10 for TIG

### Gloves

- MIG gloves: thick
- TIG gloves: thin goatskin
- Stick gloves: heavy leather

### Clothing

- 100% cotton or leather
- Welding jacket or sleeves
- No synthetics

### Boots

- Leather
- No mesh
- Steel toe optional

## 5 — GAS CYLINDERS & REGULATORS

### Common Gases

- **75/25** — MIG steel
- **100% $CO_2$** — deep penetration
- **98/2** — stainless
- **100% Argon** — TIG & aluminum
- **Helium mixes** — thick aluminum

### Cylinder Sizes

| Size | Name | Use |
|---|---|---|
| 40 cf | Small | Portable TIG/MIG |
| 80 cf | Medium | Home shops |
| 125 cf | Standard | Most common |
| 250 cf | Large | Fabrication shops |

### Regulator Tips

- Keep threads clean
- Crack valve before attaching
- Never oil regulators
- Store cylinders upright

## 6 — SHOP LAYOUT

A good welding shop is built around **workflow**, not just tools.

**Ideal Layout Zones**

1. **Cutting Zone**
   - Chop saw
   - Band saw
   - Plasma table
   - Grinding station
2. **Fit-Up Zone**
   - Welding table
   - Clamps
   - Squares
   - Magnets
3. **Welding Zone**
   - Machines
   - Gas cylinders
   - PPE
   - Fire extinguisher
4. **Finishing Zone**
   - Grinders
   - Sanders
   - Wire wheels

5. **Storage Zone**
    - Rod oven (for 7018)
    - Wire spools
    - Tungsten
    - Abrasives
    - Hardware bins

## 7 — ELECTRICAL & POWER SETUP

### Power Requirements

- 120V — light MIG, small TIG
- 240V — full MIG, TIG, Stick
- 50A breaker — most welders
- 100A sub-panel — ideal for shops

### Extension Cords

- Use **8-gauge** for 240V
- Keep cords short
- Avoid coiling (heat buildup)

## 8 — VENTILATION & FUME CONTROL

### Ventilation Options

- Fume extractor
- Overhead fan
- Cross-flow ventilation
- Downdraft table

### Best Practices

- Keep head out of plume
- Use fans to move air *across*, not *toward* the weld
- Stainless requires extra ventilation

## 9 — STORAGE & ORGANIZATION

### Rod Storage

- 7018 rods: keep in rod oven
- 6010/6011: ambient storage OK
- TIG rods: tubes or racks

### Wire Storage

- Keep MIG wire sealed
- Avoid humidity
- Store aluminum wire in dry cabinet

**Tool Storage**

- Shadow boards
- Drawer organizers
- Magnetic tool strips

## 10 — SHOP SAFETY CHECKLIST

- Fire extinguisher within 10 ft
- No flammables near welding area
- Proper ventilation
- Ground clamps in good condition
- PPE worn at all times
- Cylinders chained upright
- Clear walkways
- First-aid kit stocked

## Index

www.ingramcontent.com/pod-product-compliance
Lightning Source LLC
LaVergne TN
LVHW022207200926
848481LV00006B/1877
*9798996646913*